NATURAL HISTORY
UNIVERSAL LIBRARY

西方博物学大系

主编：江晓原

INDIAN SPORTING BIRDS

被捕杀的印度鸟类

[英] 弗兰克 · 芬恩 著

华东师范大学出版社

图书在版编目(CIP)数据

被捕杀的印度鸟类 = Indian sporting birds : 英文 / (英) 弗兰克 · 芬恩 (Frank Finn)著.
— 上海 : 华东师范大学出版社, 2018
(寰宇文献)
ISBN 978-7-5675-7992-7

Ⅰ. ①被… Ⅱ. ①弗… Ⅲ. ①鸟类-印度-英文
Ⅳ. ①Q959.223

中国版本图书馆CIP数据核字(2018)第156663号

被捕杀的印度鸟类
Indian sporting birds
(英) 弗兰克 · 芬恩 (Frank Finn)

特约策划 黄曙辉 徐 辰
责任编辑 庞 坚
特约编辑 许 倩
装帧设计 刘怡霖

出版发行 华东师范大学出版社
社　　址 上海市中山北路3663号 邮编 200062
网　　址 www.ecnupress.com.cn
电　　话 021-60821666 行政传真 021-62572105
客服电话 021-62865537
门市（邮购）电话 021-62869887
地　　址 上海市中山北路3663号华东师范大学校内先锋路口
网　　店 http://hdsdcbs.tmall.com/

印 刷 者 虎彩印艺股份有限公司
开　　本 787 × 1092 16开
印　　张 31.5
版　　次 2018年8月第1版
印　　次 2018年8月第1次
书　　号 ISBN 978-7-5675-7992-7
定　　价 880.00元 (精装全一册)

出 版 人 王 焰

《西方博物学大系》总序

江晓原

《西方博物学大系》收录博物学著作超过一百种，时间跨度为15世纪至1919年，作者分布于16个国家，写作语种有英语、法语、拉丁语、德语、弗莱芒语等，涉及对象包括植物、昆虫、软体动物、两栖动物、爬行动物、哺乳动物、鸟类和人类等，西方博物学史上的经典著作大备于此编。

中西方"博物"传统及观念之异同

今天中文里的"博物学"一词，学者们认为对应的英语词汇是Natural History，考其本义，在中国传统文化中并无现成对应词汇。在中国传统文化中原有"博物"一词，与"自然史"当然并不精确相同，甚至还有着相当大的区别，但是在"搜集自然界的物品"这种最原始的意义上，两者确实也大有相通之处，故以"博物学"对译Natural History一词，大体仍属可取，而且已被广泛接受。

已故科学史前辈刘祖慰教授尝言：古代中国人处理知识，如开中药铺，有数十上百小抽屉，将百药分门别类放入其中，即心安矣。刘教授言此，其辞若有憾焉——认为中国人不致力于寻求世界"所以然之理"，故不如西方之分析传统优越。然而古代中国人这种处理知识的风格，正与西方的博物学相通。

与此相对，西方的分析传统致力于探求各种现象和物体之间的相互关系，试图以此解释宇宙运行的原因。自古希腊开始，西方哲人即孜孜不倦建构各种几何模型，欲用以说明宇宙如何运行，其中最典型的代表，即为托勒密（Ptolemy）的宇宙体系。

比较两者，差别即在于：古代中国人主要关心外部世界"如何"运行，而以希腊为源头的西方知识传统（西方并非没有别的知识传统，只是未能光大而已）更关心世界"为何"如此运行。在线性发展无限进步的科学主义观念体系中，我们习惯于认为“为何”是在解决了“如何”之后的更高境界，故西方的分析传统比中国的传统更高明。

然而考之古代实际情形，如此简单的优劣结论未必能够成立。例如以天文学言之，古代东西方世界天文学的终极问题是共同的：给定任意地点和时刻，计算出太阳、月亮和五大行星（七政）的位置。古代中国人虽不致力于建立几何模型去解释七政“为何”如此运行，但他们用抽象的周期叠加（古代巴比伦也使用类似方法），同样能在足够高的精度上计算并预报任意给定地点和时刻的七政位置。而通过持续观察天象变化以统计、收集各种天象周期，同样可视之为富有博物学色彩的活动。

还有一点需要注意：虽然我们已经接受了用“博物学”来对译 Natural History，但中国的博物传统，确实和西方的博物学有一个重大差别——即中国的博物传统是可以容纳怪力乱神的，而西方的博物学基本上没有怪力乱神的位置。

古代中国人的博物传统不限于“多识于鸟兽草木之名”。体现此种传统的典型著作，首推晋代张华《博物志》一书。书名“博物”，其义尽显。此书从内容到分类，无不充分体现它作为中国博物传统的代表资格。

《博物志》中内容，大致可分为五类：一、山川地理知识；二、奇禽异兽描述；三、古代神话材料；四、历史人物传说；五、神仙方伎故事。这五大类，完全符合中国文化中的博物传统，深合中国古代博物传统之旨。第一类，其中涉及宇宙学说，甚至还有“地动”思想，故为科学史家所重视。第二类，其中甚至出现了中国古代长期流传的“守宫砂”传说的早期文献：相传守宫砂点在处女胳膊上，永不褪色，只有性交之后才会自动消失。第三类，古代神话传说，其中甚至包括可猜想为现代“连体人”的记载。第四类，各种著名历史人物，比如三位著名刺客的传说，此三名刺客及所刺对象，历史上皆实有其人。第五类，包括各种古代方术传说，比如中国古代房中养生学说，房中术史上的传说人物之一“青牛道士封君达”等等。前两类与西方的博物学较为接近，但每一类都会带怪力乱神色彩。

“所有的科学不是物理学就是集邮”

在许多人心目中，画画花草图案，做做昆虫标本，拍拍植物照片，这类博物学活动，和精密的数理科学，比如天文学、物理学等等，那是无法同日而语的。博物学显得那么的初级、简单，甚至幼稚。这种观念，实际上是将“数理程度”作为唯一的标尺，用来衡量一切知识。但凡能够使用数学工具来描述的，或能够进行物理实验的，那就是“硬”科学。使用的数学工具越高深越复杂，似乎就越“硬”；物理实验设备越庞大，花费的金钱越多，似乎就越“高端”、越“先进”……

这样的观念，当然带着浓厚的“物理学沙文主义”色彩，在很多情况下是不正确的。而实际上，即使我们暂且同意上述“物理学沙文主义”的观念，博物学的“科学地位”也仍然可以保住。作为一个学天体物理专业出身，因而经常徜徉在“物理学沙文主义”幻影之下的人，我很乐意指出这样一个事实：现代天文学家们的研究工作中，仍然有绘制星图，编制星表，以及为此进行的巡天观测等等活动，这些活动和博物学家“寻花问柳”，绘制植物或昆虫图谱，本质上是完全一致的。

这里我们不妨重温物理学家卢瑟福（Ernest Rutherford）的金句：“所有的科学不是物理学就是集邮（All science is either physics or stamp collecting）。”卢瑟福的这个金句堪称“物理学沙文主义”的极致，连天文学也没被他放在眼里。不过，按照中国传统的“博物”理念，集邮毫无疑问应该是博物学的一部分——尽管古代并没有邮票。卢瑟福的金句也可以从另一个角度来解读：既然在卢瑟福眼里天文学和博物学都只是“集邮”，那岂不就可以将博物学和天文学相提并论了？

如果我们摆脱了科学主义的语境，则西方模式的优越性将进一步被消解。例如，按照霍金（Stephen Hawking）在《大设计》（*The Grand Design*）中的意见，他所认同的是一种“依赖模型的实在论（model-dependent realism）”，即“不存在与图像或理论无关的实在性概念（There is no picture- or theory-independent concept of reality）”。在这样的认识中，我们以前所坚信的外部世界的客观性，已经不复存在。既然几何模型只不过是对外部世界图像的人为建构，则古代中国人干脆放弃这种建构直奔应用（毕竟在实际应用中我们只需要知道七政“如何”运行），又有何不可？

传说中的“神农尝百草”故事，也可以在类似意义下得到新的解读：“尝百草”当然是富有博物学色彩的活动，神农通过这一活动，得知哪些草能够治病，哪些不能，然而在这个传说中，神农显然没有致力于解释“为何”某些草能够治病而另一些则不能，更不会去建立“模型”以说明之。

“帝国科学”的原罪

今日学者有倡言“博物学复兴”者，用意可有多种，诸如缓解压力、亲近自然、保护环境、绿色生活、可持续发展、科学主义解毒剂等等，皆属美善。编印《西方博物学大系》也是意欲为“博物学复兴”添一助力。

然而，对于这些博物学著作，有一点似乎从未见学者指出过，而鄙意以为，当我们披阅把玩欣赏这些著作时，意识到这一点是必须的。

这百余种著作的时间跨度为15世纪至1919年，注意这个时间跨度，正是西方列强“帝国科学”大行其道的时代。遥想当年，帝国的科学家们乘上帝国的军舰——达尔文在皇家海军“小猎犬号”上就是这样的场景之一，前往那些已经成为帝国的殖民地或还未成为殖民地的“未开化”的遥远地方，通常都是踌躇满志、充满优越感的。

作为一个典型的例子，英国学者法拉在（Patricia Fara）《性、植物学与帝国：林奈与班克斯》（*Sex, Botany and Empire, The Story of Carl Linnaeus and Joseph Banks*）一书中讲述了英国植物学家班克斯（Joseph Banks）的故事。1768年8月15日，班克斯告别未婚妻，登上了澳大利亚军舰“奋进号”。此次“奋进号”的远航是受英国海军部和皇家学会资助，目的是前往南太平洋的塔希提岛（Tahiti，法属海外自治领，另一个常见的译名是“大溪地”）观测一次比较罕见的金星凌日。舰长库克（James Cook）是西方殖民史上最著名的舰长之一，多次远航探险，开拓海外殖民地。他还被认为是澳大利亚和夏威夷群岛的“发现”者，如今以他命名的群岛、海峡、山峰等不胜枚举。

当“奋进号”停靠塔希提岛时，班克斯一下就被当地美丽的土著女性迷昏了，他在她们的温柔乡里纵情狂欢，连库克舰长都看不下去了，“道德愤怒情绪偷偷溜进了他的日志当中，他发现自己根本不可能不去批评所见到的滥交行为”，而班克斯纵欲到了“连嫖妓都毫无激情”的地步——这是别人讽刺班克斯的说法，因为对于那时常年航行于茫茫大海上的男性来说，上岸嫖妓通常是一项能够唤起“激情”的活动。

而在“帝国科学”的宏大叙事中，科学家的私德是无关紧要的，人们关注的是科学家做出的科学发现。所以，尽管一面是班克斯在塔希提岛纵欲滥交，一面是他留在故乡的未婚妻正泪眼婆娑地“为远去的心上人绣织背心”，这样典型的“渣男”行径要是放在今天，非被互联网上的口水淹死不可，但是“班克斯很快从他们的分离之苦中走了出来，在外近三年，他活得倒十分滋润”。

法拉不无讽刺地指出了“帝国科学”的实质：“班克斯接管了当地的女性和植物，而库克则保护了大英帝国在太平洋上的殖民地。”甚至对班克斯的植物学本身也调侃了一番：“即使是植物学方面的科学术语也充满了性指涉。……这个体系主要依靠花朵之中雌雄生殖器官的数量来进行分类。”据说“要保护年轻妇女不受植物学教育的浸染，他们严令禁止各种各样的植物采集探险活动。”这简直就是将植物学看成一种“涉黄”的淫秽色情活动了。

在意识形态强烈影响着我们学术话语的时代，上面的故事通常是这样被描述的：库克舰长的“奋进号”军舰对殖民地和尚未成为殖民地的那些地方的所谓“访问”，其实是殖民者耀武扬威的侵略，搭载着达尔文的“小猎犬号”军舰也是同样行径；班克斯和当地女性的纵欲狂欢，当然是殖民者对土著妇女令人发指的蹂躏；即使是他采集当地植物标本的“科学考察”，也可以视为殖民者“窃取当地经济情报”的罪恶行为。

后来改革开放，上面那种意识形态话语被抛弃了，但似乎又走向了另一个极端，完全忘记或有意回避殖民者和帝国主义这个层面，只歌颂这些军舰上的科学家的伟大发现和成就，例如达尔文随着“小猎犬号”的航行，早已成为一曲祥和优美的科学颂歌。

其实达尔文也未能免俗，他在远航中也乐意与土著女性打打交道，当然他没有像班克斯那样滥情纵欲。在达尔文为“小猎犬号”远航写的《环球游记》中，我们读到：“回程途中我们遇到一群黑人姑娘在聚会，……我们笑着看了很久，还给了她们一些钱，这着实令她们欣喜一番，拿着钱尖声大笑起来，很远还能听到那愉悦的笑声。”

有趣的是，在班克斯在塔希提岛纵欲六十多年后，达尔文随着“小猎犬号”也来到了塔希提岛，岛上的土著女性同样引起了达尔文的注意，在《环球游记》中他写道：“我对这里妇女的外貌感到有些失望，然而她们却很爱美，把一朵白花或者红花戴在脑后的髮髻上……”接着他以居高临下的笔调描述了当地女性的几种发饰。

用今天的眼光来看，这些在别的民族土地上采集植物动物标本、测量地质水文数据等等的“科学考察”行为，有没有合法性问题？有没有侵犯主权的问题？这些行为得到当地人的同意了吗？当地人知道这些行为的性质和意义吗？他们有知情权吗？……这些问题，在今天的国际交往中，确实都是存在的。

也许有人会为这些帝国科学家辩解说：那时当地土著尚在未开化或半开化状态中，他们哪有“国家主权”的意识啊？他们也没有制止帝国科学家的考察活动啊？但是，这样的辩解是无法成立的。

姑不论当地土著当时究竟有没有试图制止帝国科学家的“科学考察”行为，现在早已不得而知，只要殖民者没有记录下来，我们通常就无法知道。况且殖民者有军舰有枪炮，土著就是想制止也无能为力。正如法拉所描述的：“在几个塔希提人被杀之后，一套行之有效的易货贸易体制建立了起来。”

即使土著因为无知而没有制止帝国科学家的“科学考察”行为，这事也很像一个成年人闯进别人的家，难道因为那家只有不懂事的小孩子，闯入者就可以随便打探那家的隐私、拿走那家的东西、甚至将那家的房屋土地据为已有吗？事实上，很多情况下殖民者就是这样干的。所以，所谓的“帝国科学”，其实是有着原罪的。

如果沿用上述比喻，现在的局面是，家家户户都不会只有不懂事的孩子了，所以任何外来者要想进行“科学探索”，他也得和这家主人达成共识，得到这家主人的允许才能够进行。即使这种共识的达成依赖于利益的交换，至少也不能单方面强加于人。

博物学在今日中国

博物学在今日中国之复兴，北京大学刘华杰教授提倡之功殊不可没。自刘教授大力提倡之后，各界人士纷纷跟进，仿佛昔日蔡锷在云南起兵反袁之“滇黔首义，薄海同钦，一檄遥传，景从恐后”光景，这当然是和博物学本身特点密切相关的。

无论在西方还是在中国，无论在过去还是在当下，为何博物学在它繁荣时尚的阶段，就会应者云集？深究起来，恐怕和博物学本身的特点有关。博物学没有复杂的理论结构，它的专业训练也相对容易，至少没有天文学、物理学那样的数理“门槛”，所以和一些数理学科相比，博物学可以有更多的自学成才者。这次编印的《西方博物学大系》，卷帙浩繁，蔚为大观，同样说明了这一点。

最后，还有一点明显的差别必须在此处强调指出：用刘华杰教授喜欢的术语来说，《西方博物学大系》所收入的百余种著作，绝大部分属于“一阶”性质的工作，即直接对博物学作出了贡献的著作。事实上，这也是它们被收入《西方博物学大系》的主要理由之一。而在中国国内目前已经相当热的博物学时尚潮流中，绝大部分已经出版的书籍，不是属于“二阶”性质（比如介绍西方的博物学成就），就是文学性的吟风咏月野草闲花。

要寻找中国当代学者在博物学方面的“一阶”著作，如果有之，以笔者之孤陋寡闻，唯有刘华杰教授的《檀岛花事——夏威夷植物日记》三卷，可以当之。这是刘教授在夏威夷群岛实地考察当地植物的成果，不仅属于直接对博物学作出贡献之作，而且至少在形式上将昔日“帝国科学”的逻辑反其道而用之，岂不快哉！

2018 年 6 月 5 日
于上海交通大学
科学史与科学文化研究院

出版说明

《被捕杀的印度鸟类》是英国鸟类学家弗兰克•芬恩（Frank Finn，1868—1932）的一部博物学著作。芬恩出生在肯特郡梅德斯通，曾就读牛津大学布雷齐诺斯学院。他的大部分学术成就基础是在日不落帝国的余晖之中，于其海外各殖民地——尤其是印度次大陆打下的。1892 年，他离开英国前往东非进行探险并收集鸟类标本，两年后被聘为加尔各答的印度博物馆第一助理，随后又升任副视导，直到 1903 年回国。他著作颇丰，从 1901 年起出版了十余部书籍，内容大都与印度鸟类有关。在那个年代，印度是英国各界人士的游乐场，捕猎野生动物是最受欢迎的运动之一。然而英国人对印度鸟类生态的了解，却鲜有能出芬恩之右者。他在印度博物馆期间除自行观察、记录鸟类生态及标本外，对很多英国上流人士随意丢弃的标本进行了抢救和重新评估、记录，恢复了这些蒙尘的珍物的研究价值。而殖民者和原住民无节制的滥捕滥杀，令很多印度鸟类逐渐灭绝，后人只能在芬恩的著作中寻找它们的身影了。

这本《被捕杀的印度鸟类》是芬恩著述中的名作。以 490 页内容和约 120 幅精美手绘插图，描绘了 19 与 20 世纪之交印度那些美丽鸟类不复重现的生活景象。今据原版影印。

INDIAN SPORTING BIRDS

GRUS ANTICONE.

INDIAN SPORTING BIRDS

BY

FRANK FINN, B.A., F.Z.S.

LATE DEPUTY SUPERINTENDENT, INDIAN MUSEUM, CALCUTTA

Author of "The Waterfowl of India and Asia," "The Game-Birds of India and Asia," "How to Know the Indian Waders," &c., &c.

WITH OVER 100 ILLUSTRATIONS FROM HUME AND MARSHALL'S "GAME-BIRDS OF INDIA, BURMA AND CEYLON"

LONDON

FRANCIS EDWARDS

83A, HIGH STREET, MARYLEBONE, W.

1915

$\frac{1}{4}$

CROSSOPTILON TIBETANUM

INTRODUCTION.

INDIA is a remarkable country in many ways, and not the least so in that, in spite of a civilization of immemorial antiquity, there remains in our Eastern dominions such a rich profusion and variety of wild life, even the large quadrupeds, so soon exterminated by civilized man, persisting in considerable numbers. This wealth of mammalian life naturally attracts the attention of sportsmen, to the exclusion, to some extent, of bird-shooting; yet the opportunities for the latter sport are just as excellent, and no country can show such an immense variety of sporting birds as India and our other Eastern provinces, while the individual abundance of the species is remarkable, when we consider that there is no such systematic war made upon their natural enemies as is the case in Europe; in fact, the "vermin," furred and feathered, work their will practically unchecked, and to them we must add a horde of reptilian villains, snakes, crocodiles, and carnivorous water-tortoises and lizards. The native also poaches light-heartedly everywhere, though it must be admitted that here he is only "getting his own back," as the feathered game is so numerous, and so little indented upon by a population largely vegetarian, that the comparative harmlessness of humanity to the feathered world is ungratefully repaid by considerable devastation of the crops in many districts.

The worst offenders in this way are geese and cranes, but ducks do a good deal of harm to the paddy, so that in shooting wild-fowl one is often doing actual good to the cultivator, as well as getting amusement and food; moreover, the majority of our wild-fowl being migrants from the northern regions where they breed, the stock is capable of being drawn on to a very large extent. The same remarks apply to the harmless and indeed beneficial snipe and golden plover, which, the former at any rate, form so great a stand-by of the sportsman everywhere, and to most of the sand-grouse, and some bustards; but care should be exercised in attacking the resident species of these groups, which need consideration as much as the typical game-birds of the pheasant family, under which also come the peafowl, jungle-fowl, tragopans, monauls, partridges, and quails. Of these the common or grey quail is our only migratory visitor, and being excessively abundant and widespread is the only bird of the family which is a real stand-by for shooting in the way that the various wild-fowl and snipes are. Rails are not usually shot, but, as they are regarded as game on the continent of Europe and in the United States, and as Hume thought them worth figuring, they are dealt with here along with their kin, the moorhens and coot. Many of these are also winter visitors.

But it is of course the pheasants and their allies that are the peculiar glory of Indian sporting birds, and though at present they play a very insignificant part in sport compared to their importance in Europe, systematic protection in the future ought ultimately to render them at least the equals of the water and marsh birds in this connection. Our Indian Empire is beyond comparison the richest of regions in these birds, and is indeed the metropolis of the family, including all the finest groups, except the turkeys, true grouse, and guinea-fowls.

Sporting birds are not only of interest to sportsmen, but to naturalists they are not surpassed in interest by any other groups on account of the frequent points of interest in their habits, and the unrivalled beauty of plumage which many of them display. The visitations of the migratory species — fluctuations of the commoner kinds and occurrences of the rarer ones—are also well worthy of scientific study, and much has been learnt in these particulars since the publication of Hume and Marshall's valuable work, which has of course been largely indented on in the present one, as have also the valuable publications of Mr. E. C. Stuart Baker and other contributors to the *Journal of the Bombay Natural History Society*.

In recording the occurrence of rare birds, it is not necessary that the sportsman should be able to prepare skins of his specimens, or that he should forgo eating them if rations are short; it is sufficient for purposes of identification if the head and a wing and foot be dried and if possible treated with some skin-preservative—in fact, for very big birds the head alone will be enough, and in the case of snipe the tail affords the best character for discrimination.

The scientific names used are those of the "Fauna of British India" bird volumes, now the standard work on general Indian ornithology, and where a species does not occur in these the naming of the "British Museum Catalogue of Birds" has been followed. Where the scientific name on the plate differs from these the fact has been indicated.[1]

FRANK FINN.

[1] ERRATA.—This has been overlooked in the case of the Bar-tailed Godwit (*Limosa rufa* on the plate); and the plate of the White-crested Kalij, referred to in the foot-note on p. 183, is not one of those in this book. On p. 205, also, *Ceriornis* should read *Tragopan*.

ANAS CARYOPHYLLACEA

LIST OF PLATES.

W. Foster.

ANAS BOSCHAS

INDIAN SPORTING BIRDS.

Mallard.

Anas boscas. *Nil-sir*, Hindustani.

Although mallard are far from being generally distributed over our Eastern Empire, as being *the* wild ducks of the Northern Hemisphere generally, and the ancestors of most of our tame ducks, they deserve to head the list of typical ducks, being also themselves the type of all and exemplifying several points which must be referred to by anyone dealing with the group.

The lovely green head, white collar, chocolate breast, curled black tail, and splendid wing-bar of blue and white are so distinctive of the mallard drake that little need be said about his plumage, which has for the most part a sober pencilled-grey coloration, beautifully setting off the brighter tints. But the female, whose plumage, as is usual in the most typical ducks, is of a mottled-brown tint, is naturally much like several others; her distinguishing mark is the blue, white-edged wing-bar which she shares with the drake. This blue ribbon-mark will distinguish her from all our ducks but the Chinese grey duck or yellow nib (*Anas zonorhyncha*), which bird has a black bill with a yellow tip, and is much greyer in tint, with a dark sooty belly. The bill of the female mallard is dull orange, with a large, dull black patch occupying most of the centre part; the drake's is a sort of sage-green, sometimes verging on yellow.

This point is worth mentioning, because when the drake goes into undress plumage the colour of his bill does not change as in some species at this time. In plumage the mallard in this "eclipse" stage is very like the female, but not exactly, the crown of the head and the lower back down to the tail being black, not streaked with brown. Young drakes assume a plumage similar to this for the first feathering, though at first sight all the brood look much alike. The undress plumage is assumed after breeding, about June, and lost about September; it comes on at the time when all the great wing-quills are moulted, so that the birds cannot fly for some weeks. This peculiarity of putting on a plumage more or less like that of the female characterizes most ducks in the Northern Hemisphere when the sexes have a very distinct plumage; it is curious that it is carried in the summer, when most birds are in the highest feather; but the facts that it is really a winter plumage in the cotton-teal, and almost so in the garganey, and that ducks as a matter of course start courting in the autumn as soon as they get their gay plumage, suggest that it is really a winter plumage that has had a tendency to be shifted earlier and earlier till it is now a summer one.

Mallard weigh in the wild state in India about two and a half to three pounds in the case of drakes, and even up to four; females are about two, and may approach three. Domestic ducks in India are not much bigger than this, though they may look so on account of their coarseness and loose feathering; as they often resemble mallard in colour it is just as well to be careful how one shoots at unusually unsuspicious-looking ducks until one is sure they really are wild.

Wild mallard in India are not as a rule to be expected away from the North-west, and even there it is only in the extreme end of that region that they are abundant; and south of Bombay they are unknown. As a straggler the mallard occurs all along our Northern Provinces as far as Mandalay; in Cachar, Mr. E. C. S. Baker reports it as "not very rare." One wants to be quite sure, however, of any given bird being really a mallard when it is shot out of the North-west Province; of course the full-plumaged drake is unmistakable, but there are several ducks very like the female, the yellow-nib even having the blue wing-bar as above noted. As mallard breed in Kashmir they often have not very far to come to get to their winter quarters in some cases, though many winter in Kashmir itself; in Sind they are very common, and it is only here that hundreds may be seen in a flock; elsewhere the parties are small, and odd specimens, where the species is rare, may be found associating with other kinds of duck.

The habits of this duck are thoroughly well known, as almost everyone, even if not a sportsman, has had ample opportunities of observing the bird in a protected state in parks. It is not highly specialized in any way, but a thoroughly robust and vigorous bird; it swims, walks, and flies, with ease and efficiency, but in no separate department equals some other species—for instance, it cannot fly so well as a gadwall, run so well as a sheldrake, or swim so fast as a pochard. It dives fairly well to save its life, or in play, but seldom does so to get food; when I have seen this done it has always been by females or young birds, never by old drakes; but the action is a rare one even with the other sex when adult. Females, however, are also said to be more cautious and cunning in concealing themselves after being wounded than males.

When a pair are together on the water, the drake waits for the duck to rise first; his note, a faint wheezing *quaykh*, is very distinct from the duck's well-known quack or rather *quaak;* but though this was pointed out by White of Selborne more than a century ago, it does not seem to have been fully realized even now that the same distinction of voice applies to a large number of the ducks, and that the two notes in these cannot be interchanged, the drake having a large bulb in the windpipe at its bifurcation towards the lungs, which absolutely modifies the sound and prevents him giving the female call, while similarly she cannot imitate his. In the breeding season the mallard drake whistles as well as wheezes, and the duck talks affectionately to him in short staccato quacks, with sidelong noddings of the head; he for his part plays up to her by rearing up with his head bent down, then dropping on the water and jerking up his stern, at the same time displaying by a slight expansion of the plumage the bar on his wings. Anyone may see these antics among domestic ducks—common ones, I mean, not Muscovy ducks, which have very different ways, and are descended from the South-American *Cairina moschata.*

Mallard, in conformity with their usual unspecialized ways, are not particular about the water they frequent so long as it affords safety during time of rest, or food when they are in search of this. Thus they will frequent small dirty ponds or big open broads; feed on land as well as in water, and by day as well as by night. In fact, I think many ducks are chiefly made nocturnal by our persecution; I rather fancy that the mandarin is really the only true night bird in the family, as he is habitually very quiet during the day even in captivity. The food of mallard is pretty nearly everything: corn, herbage, roots, worms, or any other small animal life, berries, acorns, &c., &c.; as long as there is plenty, they are not particular. They are themselves almost always excellent, at any rate in India.

They breed in Kashmir, in May and June, making a well-concealed, down-lined nest among ground cover, as a rule; now and then among water-plants, rarely in trees; the eggs are usually eleven and their grey-green colour is well known. The ducklings are clad in black and yellow down. The Indian native names, besides *Nilsir*, are *Lilg* in Nepal, with the female form *Lilgahi*.

Spotted-bill.

Anas pœcilorhyncha. *Garm-pai*, Hindustani.

The spotted-bill might well be called the Indian mallard; it is so like the female of that bird, or rather perhaps like some abnormally coloured tame duck, that it would hardly attract attention at a distance, the only conspicuous colour point being the broad snow-white streak along the sides of the dark hinder back, which streak is the outer webs of the innermost wing-feathers. Close at hand, several detailed differences become noticeable; the brilliant and characteristic coloration of the bill, twin-spotted with scarlet at the forehead, jet-black in the middle, and rich yellow at the tip; the bright green instead of blue wing-bar; and the way in which the plumage, pale drab speckled with black in the forequarters, gradually shades into black at the stern, a style of coloration never matched among all the many varieties of the tame duck.

In young spotted-bills the characteristics of the species are not so well developed; the colours of the beak not being

ANAS PÆCILORHYNCHA

separated into the definite tricolour in many instances, the base being orange and the sides as well as the tip yellow. When it comes to the voice, the relationship of the spotted-bill and mallard is again obvious at once ; in both the quack of the duck and the wheeze of the drake are the same, although the latter in the spotted-bill bears the same unpretentious plumage as his mate.

In weight spotted-bill are pretty much the same as mallard ; there is, perhaps, not quite so much difference in the size of the sexes in the Indian bird, and the male spotted-bill does not run so heavy as some mallard—it is a noticeably lighter-built bird when the two are closely compared in life.

The spotted-bill inhabits nearly all our Empire, but is not found in Southern Burma or the islands of the Bay of Bengal ; nor does it ascend the hills higher than about 4,000 ft. It never leaves our limits entirely, but, like all birds whose livelihood depends on water, has to shift its quarters more or less to secure favourable conditions. In Central India and in Manipur it is more common than anywhere else.

It is not very particular about its haunts, frequenting small or large ponds, running or standing water ; but on the whole standing water with plenty of cover is most to its taste. It does not associate in large flocks like its migratory allies, and pairs are commonly found ; a solitary bird will sometimes assume the honorary headship of a flock of teal, but they keep apart from other waterfowl as a rule. On a few occasions as many as a hundred spotted-bill have been seen in a flock, but half this number is rare, and small flocks of about a dozen are usually seen.

The general habits of the spotted-bill are so exactly like those of mallard that it is no wonder the two are sometimes confused. Like its migratory cousin, the Indian bird flies, swims, walks, and dives well ; although it rises with more of a fluster and does not get up its pace so quickly, it has the advantage when wounded, as it dives very well and hides most cunningly in any available cover. Its tastes are as omnivorous as those of mallard, and it is a pest to rice-growers at times. Even its nesting-habits are similar, as it breeds on the ground in grass or other shelter, not in the elevated sites usually favoured by most of our resident ducks. The eggs can be distinguished from mallard eggs by their rounder form and more buff tinge, not being greenish; ten is the usual sitting, but the full number of ducklings are never apparently reared; in fact, considering the ground-breeding habits of the bird, and the abundance of all sorts of vermin in India, the wonder is that it is so common at all. It is universally valued as a sporting bird, and is as good eating as mallard; so close is the alliance that the two species when brought together in captivity interbreed without hesitation. The native names, however, keep up the distinction of species; the spotted-bill being called *Hunjur* in Sindhi, and *Gugral* as well as *Garmpai* in Hindustani; *Kara* is the Manipuri name, and *Naddun* that used in Nepaul.

Yellow-nib or Chinese Grey Duck.

Anas zonorhyncha.

This East Asiatic representative of the spotted-bill has of late years proved to occur quite commonly is Assam, and also to be found in the Shan States and Upper Burmah. Although it never has the red spots at the base of the bill so characteristic of the spotted-bill of India, this is not the chief distinction, as these spots may be absent in perfectly adult and otherwise normal spotted-bills; I have seen one such in the London "Zoo," which was the father of perfectly normal young when paired to a female which showed the spots.

The most striking point about the yellow-nib is the fact that the wing-bar is blue instead of green, and that the white on the inner quills never forms more than a border, and does not take up all the outer half of the feather; moreover, the plumage is not so distinctly marked as that of the spotted-bill—although there is a very distinct whitish eyebrow—and is much darker below, the belly as well as the stern being blackish. The bill is apparently rather smaller, and is blacker, the yellow tip being smaller.

The yellow-nib occurs sometimes in quite large flocks—as many as forty have been seen together in Lakhimpur—but small parties of pairs are commoner; when thus few in numbers they

ANAS ♂ STREPERA

associate with other surface-feeding ducks; they are wild, like ducks in Lakhimpur generally.

The eggs have been taken in Dibrugarh and the Shan States, and are like spotted-bills' eggs, but run smaller. The best known haunts of this bird are from eastern Mongolia east and through China and Japan to the Kuriles; no doubt in the northern part of its range it is migratory, and it would appear to have a longer wing than our resident Indian spotted-bill.

Gadwall.

* *Chaulelasmus streperus.* *Mila*, Hindustani.

Not at all a familiar bird at home, the gadwall is, in the East, the most abundant of all the larger winter ducks, and holds much the same place in shooting in most districts as mallard in the west. The female is very like the female mallard, having a similar mottled-brown plumage, but the bar on the wing is all white and there is generally a little chestnut in front of it. The drake is a very poor creature compared to the splendid mallard drake; his head is of a dull speckly brown and the pencilled grey of his body is dark and dull in tone, the only striking note of colour being the velvet-black stern. The white wing-bar is preceded by a patch of reddish-chocolate.

Plain as his plumage is, the gadwall drake yet changes it for female dress, like the mallard, in the summer; indeed he goes further and changes his black bill for the orange-edged one of the female, but the distinct chocolate patch on the wing remains to distinguish him. Young males have less of this.

Gadwall are finer-boned and more delicately framed ducks than mallards, and are not quite so large, although they have a plump appearance; the drake seldom weighs over two pounds, and the duck does not reach that weight, and may be as little as one; they are generally in good condition, even when recently arrived, at which time ducks are apt to be poor after their exertions in the long flight.

Although they penetrate to most parts of the Empire, they

---

* *Anas* on plate.

are not so widely distributed as the pintail or even the humble shoveller, to say nothing of the teal, for they do not visit the extreme south of India nor Ceylon, to say nothing of the islands of the Bay. They come in about November and may stay on as late as May, though March is the more usual month for their departure. They are found in flocks of various sizes, and are not naturally remarkable either for shyness or its opposite, though after persecution they give trouble enough to the gunner. Like mallard they rise smartly, and their flight is more rapid, and somewhat teal-like both in style and sound. They sit rather high in the water, and swim and walk with ordinary ability, not infrequently coming ashore to feed; among the items there sought for Hume enumerates small moths and butterflies—rather ethereal diet for a duck one would think, especially as so few birds have been actually observed eating butterflies at all. Water-insects and shell-fish are also partaken of, but the gadwall is mainly a vegetarian feeder, especially appreciating wild rice and paddy, even when half ripe. It is almost always an excellent bird for the table.

Gadwall like a certain amount of cover in the water they frequent, but are not particular birds about their habitat; their visits to the rice-fields are made in the mornings and evenings, and by day they retire to the broader waters to rest. They seem confined to fresh water. Although not bad divers when urged by necessity, they do not seem to dive for food; this, of course, is what one would expect, but there exists an old statement to the effect that the gadwall dived freely and frequently; this was probably founded on observation of some unusually gifted individual bird. The gadwall's quack is more shrill than that of the mallard, and weaker and sharper, and more often used, according to Hume. This presumably refers to the female; the male, which is pretty noisy towards spring, has a gruff, grunting quack, not at all like a mallard drake's note, or indeed like that of most male ducks, though the shoveller and clucking-teal have voices of somewhat the same type.

Gadwall are not only thought highly of by sportsmen, but seem to be popular in pond society; they are found, according to Hume, in the company of all sorts of other ducks, and are

ANAS ⅓ FALCATA

even tolerated by geese, who usually maintain an attitude of disagreeable exclusiveness. The gadwall drake shows off to his mate in the same attitudes as the mallard, at which time the sudden exposure of his snow-white wing-bar has a curious flashing effect against the dull iron-grey plumage. In captivity he is a devoted mate, vigorously defending his chosen duck, and strictly refraining from aggression towards the mates of his neighbours, a virtue not so common among ducks of the mallard group as it might be. He is, in fact, the typical, good, steady, reliable bird of his tribe. Outside India the gadwall has the same wide range all round the northern parts of the world as the mallard, but it is not always equally common everywhere; it is not, for instance, abundant at the very far end of Asia any more than in Britain. The Bengali name is *Peing-hans,* the Nepaulese *Mail,* the Sindhi *Burd;* while other appellations are the Hindustani *Bhuar* and *Beykhur.*

Bronze-capped Duck.

* *Eunetta falcata.* *Kala Sinkhar*, Hindustani.

I quite agree with Hume that the name of "falcated," applied generally to this bird, is not English and is misleading; but I cannot follow him in calling it a teal, for its size is so much larger than that of any teal, and its affinities to the gadwall so obvious, that that term is misleading also. Indeed, the female is almost exactly like the gadwall female, though easy enough to recognize if one remembers that it has the wing-bar black instead of white, the feet grey instead of orange, and the bill all black, not orange-bordered.

The full-plumaged drake, with his lovely silky-crested head of green and bronze, his white neck crossed by a dark-green collar, and the bunch of long curved feathers in each wing—it is these that are "falcated" or sickle-shaped, not the *bird*—is at once distinguishable from all other ducks. The general plumage is grey at a distance, but close at hand is seen to be made up of pencillings of black and white, as in most grey-looking ducks. The

* *Anas* on plate.

black and yellow tail-coverts conceal the tail, and give the bird a very stump-ended look; in fact, in life it is not so beautiful as artists make it, but looks thick-headed and top-heavy, lovely as its plumage looks in the dead specimen. It is also, in captivity at any rate, very quiet and uninteresting.

The weight of a male is about a pound and a half; his note is a whistle, while the duck has the ordinary *quack*, five times repeated.

This bird breeds in Siberia and winters in Japan and south-eastern Asia, including India; as Hume very accurately suggested it might be, it certainly is commoner in India than the clucking teal, a bird of similar range, although it is certainly one of our rarities still. Hume had got no less than five specimens of it by the time he published his account in the "Game-birds" at the end of the seventies, nearly all of them in Oudh; but it has since been found further east, as far as Upper Burma and Manipur. The Calcutta Bazaar is a good place to get it; one of Hume's five came from there, and from 1897 onwards for the next four or five years I never missed seeing it, and in 1900 and 1901 it could fairly be called common. I have seen a dozen or more in a good season, but I should say not twenty-five per cent. were males. The male in undress, by the way, is almost exactly like the female, but has the inner quills black and grey, rather like their colouring, when long and curved, in the full plumage; the true sickle feathers, like the Mandarin's fans, do not appear till the rest of the plumage is perfect. Fresh-caught birds are very wild, and the species is said to be a strong flyer.

Pintail.

Dafila acuta. *Sink-par*, Hindustani.

The elegant clipper-built pintail is at once conspicuous by his racing lines among all our ducks, his long neck and long sharp tail making him conspicuous either in the air or on the water. His colouring is chiefly remarkable for the large amount of white, this reaching below from the liver-brown head to the black stern; the upper-parts are of the finely lined grey so common among the males of the duck tribe.

♂

DAFILA ACUTA.

The female, having a much shorter, though still pointed, tail, and the usual mottled brown plumage of typical ducks of her sex, is less easily recognizable, but she is recognizable on close inspection by having the tail feathers marked with light and dark cross-bars, instead of light-edged and dark-centred as usual, and by having either no wing-bar at all or one, like the drake's, of the unusual tint of bronze.

The drake in undress has at first a very feminine aspect, but his tail, though short, is still darker than the female's, and his plumage is rather cross-barred than mottled with curved markings.

Pintail weigh about a couple of pounds in the case of drakes, ducks being about half a pound less.

They are among the most valued sporting birds in India, coming in vast numbers every winter, and spreading all over the Empire ; the flocks are seldom under twenty in number, and generally contain two or three times that number of birds, while hundreds and even thousands may be found together. Although so sociable, in some cases flocks may be found consisting of drakes alone. They like large pieces of open water with plenty of surface weed for their day-time rest, and keep a good look-out, being naturally wary ; they do not move till well on in the evening, and then go out to feed in all sorts of watery places, returning at daybreak to their resting-places. They fly, as might be expected, from their slender shape, with very great speed, and are considered to be the swiftest of all the tribe. The sound of a flock passing is described by Hume as a "low, soft, hissing swish," which is quite unmistakable. On the other hand, their swimming and walking powers are but ordinary, and they dive badly. Their long necks are of great service to them in feeding on the bottom with the stern up, and also in reaching up to pull down paddy-ears, for they do not disdain vegetable food, although chiefly animal feeders, especially eating shell-fish ; this, however, does not give them an unpleasant flavour, and, as a matter of fact, few ducks are so uniformly good.

In general disposition they are placid, rather characterless birds ; they are not even noisy, the drake's note being singularly soft and subdued, and very hard to describe ; the duck's quack is harsher than that of the mallard or spotted-bill, but she very seldom utters it. The drake when courting shows off like the mallard, but is rarely seen to do so. In fact, though as a sporting bird the pintail is unrivalled among the ducks, and has few equals among other groups, from the naturalist's point of view he is disappointing, in spite of his elegant and refined appearance.

Pintail, like mallard, are found all round the world, but only breed in high northern latitudes as a rule; wild-bred hybrids with the mallard sometimes occur, but such have never turned up in India. These much resemble rather delicately shaped and tinted mallard, but have the tail only curved, not curled, the head less richly glossed, and the breast light-fawn, not chocolate. This hybrid, by the way, is quite fertile in captivity.

In Bengal pintail are called *Dighans* or *Sho-lon-cho;* in Sind *Kokarali* or *Drighush; Digunch* in Nepal and *Laitunga* in Manipur, while another Hindustani name besides *Sink-par* is *Sank.*

Shoveller.

Spatula clypeata. *Tidari*, Hindustani.

The long and broad-tipped bill of the shoveller, very like a shoeing-horn in shape, and provided along the edges with a comb of horny sifting-plates, is so characteristic that anyone could pick the owner out in the dark by merely feeling its beak. It is therefore unnecessary to go into any details about the plumage of the mottled-brown female, but in justice to the drake it must be mentioned that he combines the mallard's green head with the pintail's white breast, and wings bluer than the garganey's with flanks and belly redder than the Brahminy.

He is, in fact, a very flashy-looking bird when in colour, but in undress plumage he is very like his brown mate, but is distinguishable by having the blue wing-patch. Even his bill at this time changes colour, from jet-black to the olive and orange of the female. He keeps his undress a long time, not coming into colour as a rule before Christmas. Take away his bill and wings, and the shoveller is a rather small duck, only weighing about a pound and a half with those appendages included.

$\frac{2}{3}$ SPATULA CLYPEATA

He is also of rather small account from a sporting point of view, for though one of our very commonest winter ducks, spreading all over the Empire except the islands in the Bay of Bengal, he is not numerous anywhere, going in small flocks or pairs, which somewhat affect the company of other species. His tastes, moreover, are low; although to be found here and there in any sort of watery environment, what he really likes is muddy shallows and weedy ponds, and even dirty little village tanks, where stores of organic matter appeal to his palate. He is exquisitely provided for extracting nutriment suspended in water by his wonderful bill, which, as Darwin long ago pointed out, is like the mouth of a whalebone whale in miniature; the principle is the same in all ducks' bills, but in the shoveller it is carried to perfection, and so this bird seldom feeds by exploring the bottom or foraging on shore; but paddles slowly about, often turning in a circle, and bibbles assiduously, finding food where no other duck could obtain it. Any sort of food, vegetable or animal, passes muster with him, but of course he is no dirtier a feeder than other ducks when found in a clean environment; he simply takes what comes first. But in any case he has a bad name as food, though I must say I think this may perhaps be exaggerated by a natural prejudice against a bird often seen in dirty places. This duck is not a wary bird, but, in spite of its lazy and slow movements on land and water, its small feet and short legs not being suited for rapid running and swimming, it is active enough on the wing, and will even oblige a flock of teal with a lead. It cannot dive much, so is easily captured if winged.

The note of the male is *quuck quuck*, but one does not often hear it except when he is courting; he is dull and stolid then as at most other times, simply moving his head up and down in a daft sort of way. The female seems to have the ordinary quack. Shovellers come into India rather late, about the beginning of November, and sometimes spend all the winter in Kashmir; they are also late in leaving, staying in some places as late as April or even May. One has been met with with a brood in Ceylon in March, but such breeding in our limits is doubtless quite exceptional. In its breeding haunts, which include the north temperate portion of the whole world, it nests on the ground and lays nine eggs or so of a yellowish-grey colour. The young are very small, hardly bigger than young teal, and their beaks are not broadened at first, though rather long, but they start surface-bibbling and revolving round and round at once.

The shoveller is well off for names; in Nepal even the sexes are distinguished, the male as *Dhobaha*, and the female as *Khikeria*, *Sankhar*; in Sind the name is *Alipat* and in Bengal *Pantamuhki*, while, in addition to that given at the head of this article, there are other Hindustani titles—*Punana*, *Tokarwalah*, and *Ghirah*.

Common or Green-winged Teal.

* *Nettium crecca*. *Lohya Kerra*, Hindustani.

The common teal, the smallest, and one of the handsomest and most sporting of our migratory ducks, can be at once distinguished from all the rest by the brilliant patch of metallic emerald green on the wing, whence the name green-winged teal often applied to it to distinguish it from the blue-winged teal or garganey. Except for this wing-mark, there is nothing distinctive about the mottled-brown plumage of the female, but the drake is a most handsomely coloured little bird, with his chestnut head widely banded with green, the cream and black stripes on his pencilled-grey back, and the thrush-like spotting on the breast. This teal, though with a proportionately long narrow bill, is a thick-set, plump little bird, weighing from 7·7 to 12 ounces, with no noticeable distinction in this respect between the sexes.

In the drake's summer undress, which he loses later than is usually the case, so that specimens bearing more or less of it are usually seen even in their winter quarters here for a month or two, he is generally like the female, but has the breast plain brown without speckling, and the markings of the body less well defined. The drake's note is a whistle, the duck's is a tiny quack; he shows off to her like the mallard, but with a quick, jerky action.

Common teal are just as familiar in the East during the cold

---

* *Querquedula* on plate.

QUERQUEDULA CRECCA

weather as they are in Europe, in fact more so ; they are some of the commonest of our migratory ducks, and are certainly the most widely diffused, being found practically all over the Empire, even penetrating to the Andamans and Nicobars, though apparently not to Southern Tenasserim. They come in early, many arriving in September, and some occasionally even before August is out; but October is, as usual with our migrants, the month for the main body to arrive in.

In the north-west, where they are most abundant, flocks of thousands may be seen, but two or three dozen is a usual figure for the flocks commonly met with, and even single birds as well as pairs often turn up. Any sort of water may hold them, as they are content with a very small area, but they like plenty of cover, and lie fairly close, so that they afford frequent shots. They swim and walk fairly well, but do not come on land much for a surface-feeding duck; their diving powers are nothing extraordinary, but they are adepts at taking cover under water when wounded, so that where there are plenty of weeds, &c., they are hard enough to get hold of.

Their flight is exceedingly fast, but like most small creatures, furred as well as feathered, they are probably credited with more speed than they actually possess, owing to the quickness of their movements, which has a deceptive effect ; at any rate, the shoveller and even the spot-bill, can give a flock of teal a lead. Their really strong point on the wing is their power of wheeling suddenly, which often proves too much both for the peregrine falcon and for the human enemy with his gun. Their feeding-time is mostly at night, and the food itself vegetable for the most part, though small live things are not despised ; but Hume argues reasonably that they must be mainly vegetarians, because in the "tealeries" so common in upper India in his day the birds throve on paddy and lucerne only, and kept their condition, if well looked after, all through the hot weather and rains, when they were much valued as food when butcher's meat was unattainable. Hume indeed considered that a well-kept captive teal was even better than a wild one, and the wild bird is universally praised for its excellent qualities ; I do not know any bird I like better myself, as there is something about it one does not get tired of. In disposition the teal is very sociable and fond of its mate; it is also excessively "cheeky" with larger ducks; I have several times seen a full-winged one which had been bred at the London "Zoo" and used to return to visit its pinioned comrades, in the thick of a fight with an Andaman teal or Chilian wigeon (*Mareca sibilatrix*), both of them far bigger and redoubtable fighters in their way, and I once saw another in St. James's Park chasing a female mallard, to her great indignation and the surprise of her mate. These teal may now and then be found in India in any month of the year, but there is no reason to believe they breed there. They are found all across the Old World. Some Hindustani names are *Chota Murghabi*, *Putari*, and *Souchuruka*; *Baigilagairi* is used in Nepal and *Kardo* in Sind; while the Canarese name is *Sorlai-haki*, the Tamil *Killowai*, and the Bengali ones *Naroib* or *Tulsia-bigri*.

Garganey.

Querquedula circia. *Chaitwa*, Hindustani.

The garganey is often called the blue-wing teal to distinguish it from the common or green-winged teal; there is, as a matter of fact, no actual bright blue about the wing, but the inner half, in the drake, is of a delicate French grey, very noticeable in flight, and his white eyebrows are also striking points; while on the water the mottled brown of the fore- and hind-parts, contrasting with the grey of the sides, are characteristic. Except for the wing-bar, which is of a rather subdued green, there is no bright colour about this little duck, but nevertheless he is a very striking bird.

The female, in her plain mottled-brown plumage, is at first sight just like the female common teal, but has not the brilliant green wing-patch. The male in undress can be distinguished by the lavender and green on the wings; on the water with wings folded he is just like his mate, and he bears his undress plumage longer than any other duck, not coming into male colour till the spring. The garganey is a slightly bigger bird than the common teal, weighing generally about thirteen ounces and even reaching

QUERQUEDULA CIRCIA

a pound. It has a rather shorter beak, and is generally more shapely and fashioned like a miniature mallard.

No duck visits us in greater numbers than this; in fact, i what one saw in the Calcutta Bazaar in the nineties was any criterion, this bird is in winter the most numerous duck in the country, surpassing even the whistler and the common teal. It habitually associates in flocks of hundreds and even thousands; parties of less than a score are uncommon. The large flocks are mostly to be found in the north-west, though the bird is distributed over India and Burma generally, and is well known in Ceylon. It has less predilection for small and weed-grown bits of water than the common teal, and is quite at home on wide lakes and rivers, where by choice it spends the day. It feeds mostly at night, and in some localities destroys the paddy by the acre, being chiefly a vegetable feeder, though of course, like ducks in general, it does not despise any animal food it comes across. In "tealeries" also, it is found to thrive on the same vegetable *régime* as the common teal; it is never, however, quite so good a bird on the table.

In disposition and style of flight it is decidedly different; it is, as a general rule, much wilder, and flies much straighter and to a greater distance when alarmed; the flocks pack very close, and as they pass overhead the sound made by their wings—a pattering swish, Mr. E. C. S. Baker calls it—is very characteristic. They swim and walk as well as common teal, and dive much better; Hume sums the matter up by saying they are more vigorous and less agile birds. Garganey are not at all noisy birds; the duck quacks, but the note of the drake is as different from that of the common teal as it can well be; it is a sort of gurgling rattle, most unmistakable when once heard. It is constantly uttered during courtship, when the bird does not rear up like the common teal, but merely moves his head up and down like the shoveller; in fact, to this bird all the teal with blue or bluish patches on the wings seem to be related. Judging from the note, it is no doubt this bird, not the common teal, that was the original *Querquedula* of the ancients—the Spanish name *Cerceta* comes very near this Latin one. Although not nearly so common in Europe as in the East, the garganey is well known there; it breeds in small numbers in England, where, unlike all our other ducks, it is a summer migrant only. It seems never to go into cold waters, though able to bear our English winters in captivity quite well. All across Asia it is to be found in summer, and is a very common species in winter from Egypt on the west to China on the east, and even reaches Java. It comes in nearly or quite as early as the common teal in India; in fact, in the north-west generally earlier, and in two instances has been found breeding in India and Burma, for though, as a rule, only a visitor, it may be found at any time during the year exceptionally. The breeding records, however —one from Oudh and one from Moulmein—only concern the capture of more or less fledged young, and an actual nest has not been found.

The nest in the countries where the bird breeds is made on the ground among grass or other cover near small pieces of water, and is provided with a rather scanty lining of down. About eight is the usual number of eggs laid; they are yellowish-white, like those of the common teal, and of the same size. The ducklings also are much like miniature young mallard.

The Bengali names of the garganey are *Gangroib* and *Girria;* in Hindustani it is called *Khaira* and *Patari* as well as *Chaitwa.*

Clucking Teal.

* *Nettium formosum.*

The head alone is quite enough to identify the drake of this rare species; the throat and crown are black, the face buff, with a black line down from each eye as if the bird had been crying tears of ink, and a crescent of glittering green curving round at the back. There is a vertical white bar on each side of the pinkish, black-spotted breast, separating it from the pencilled blue-grey flanks, and the shoulders are decked with long hackles streaked with black, buff, and chestnut.

---

* *Querquedula glocitans* on plate.

W. Foster.

F. Wailer Chromo Lith. 15 Hatton Garden London

QUERQUEDULA GLOCITANS.

The female is much like that of the common teal, but is larger, with much shorter bill in proportion, and a very distinct white patch at its base; moreover, the wing bar is mostly black and white, with only a narrow streak of green, and this running vertically parallel with the white border, not longitudinally. The male in undress differs from her in having the lower back plain brown as in the full plumage, not mottled. For some time after he loses this eclipse dress in the autumn his beautiful head markings are much obscured by brown edgings to the feathers, although the strange pattern is quite recognizable.

This teal is not only larger than the common and garganey teals, longer-tailed and shorter-billed, but stands much higher on the legs, and runs very actively. The loud clucking note of the drake, which sounds like *mok-mok*, is most characteristic, and the bird can never hold his tongue for long. He displays in a curious way, generally on land so far as I have seen; first raising his head and erecting the plumage on it, so that it seems much larger, and then jerking it back on to his shoulders, clucking vigorously the while.

The clucking teal is an eastern Asiatic bird, for though breeding freely in Siberia and sometimes occurring to the westward in Europe, its chief winter haunts are Japan and China, where it must be extremely abundant, judging from the thousands of live birds that have lately been exported to Europe and even Australia of late years. At the time of writing it is hardly dearer in England than common teal, and the dealers have scores at a time. Less than a dozen specimens have been taken in India, and these chiefly of recent years; but one was got in the Calcutta Market in 1844. I also got the first female recorded there. No doubt this sex has been often overlooked, for all the other records seem to be of males, and these have been got as far apart as Gujarat and Dibrugarh.

Marbled Teal.

* *Marmaronetta angustirostris.*

The Marbled Teal is decidedly large for a teal, weighing about a pound. Its peculiar colouring, which is the same in both sexes, is very distinctive, though also very unpretentious. It is a mottling of drab and cream-colour, the only approach to distinct and easily apprehensible colour-points being an ill-defined dark eye-patch and pale grey edgings to the wings. It has a very faded and washed-out appearance, but its dark slaty bill and feet contrast strikingly with the pallid plumage.

The pale or cinnamon variety of the garganey which sometimes occurs has been mistaken for this bird, and so has the female mandarin duck; but neither has the dark eye-patch, and the mandarin is not mottled on the back, while the garganey is far smaller than the present species, and when albinistic, generally has the beak and feet flesh-coloured.

The marbled teal, as a rule, is only found in winter, and chiefly in the North-West Provinces, extending as far as Oudh; it has even strayed to the neighbourhood of Calcutta, but it is only really common in Sind, where it may fairly be called abundant. It is among the surface-feeders what the white-eyed pochard is among the divers, a bird of coot-like proclivities, preferring water with plenty of rushes growing in it, and getting up, not in flocks but independently like quail; when on the wing also, they fly low and not very fast. They will dive and hide under water with the bill out when wounded, and seem seldom to come ashore, though they walk well, as might be expected from their light build, which they share with the Andaman and clucking teals, both good pedestrians. When courting the drake jerks back his head on to his shoulders; his note is the usual whistle of teal drakes, while the duck quacks. It feeds pretty equally on both vegetable and animal food, and again like the white-eye, is not a good table bird.

Some eggs found in a nest under a babul bush in a salt marsh

---

* *Querquedula* on plate.

$\frac{1}{2}$

QUERQUEDULA ANGUSTIROSTRIS.

on the Mekran coast are supposed to be those of this bird ; they were taken on June 19. The certainly known eggs are cream colour, as were these. The ordinary range of the species is from the Canaries and Cape Verd Islands, along the Mediterranean region, and through Western Asia, so that in range, as well as colour, it rather recalls a sand-grouse.

Andaman Teal.

* *Nettium albigulare.*

The Andaman teal, although not distinguished by Hume from the Australasian oceanic teal (the real *gibberifrons*) has, with the exception of a single specimen recorded from Burma, no doubt a wind-blown straggler, been only found in the Andaman Islands.

It would, however, be easily recognizable among our mainland species, owing to its very dark brown colour, but little relieved by the pale edgings to the feathers, and the conspicuous white patch on the wing in front of the wing-bar, which marking is black, green, and white, the two last colours forming narrow central and bordering streaks respectively. Except in yearling birds, there is a white ring round the eye, and some older specimens have the feathers at the base of the beak, or even the whole sides of the head and back of the neck white. This amount of white colour takes some years to develop, and I have only seen one bird wild killed showing it, but it has appeared sooner or later in all captives I have watched, though those bred in captivity are without even any white eye-ring at first; just like young wild birds in fact. It remains to be seen whether this marking is really becoming common among the wild stock from some cause we are unaware of, or whether only captive birds get a chance to live long enough to become white-headed, for that is what the tame ones ultimately become, the brown being limited to the centre of the crown.

The Andaman teal resembles the small whistler in many of its ways, being active on the water, a regular percher and a light

---

* *Mareca gibberifrons* on plate.

flyer, and an inveterate fighter; this it has to be if it must live along with the Andaman strain of the small whistler, which, judging from a pair we had in the Calcutta Zoo, is even more peppery in character than the mainland birds, from which these were distinguishable by their smaller size and richer colour. Like the whistling teal, also, it breeds either in trees or on the ground, the eggs being cream-coloured; the tree site for the nest is the common one, and it is placed in a hole. It is, however, a true teal in general characters, though large for such a bird, weighing about a pound; also, though the sexes are alike in colour, it has the usual sex difference in note found in the teal, the drake whistling and the duck quacking. It is a very active runner, and also flies sharply though noiselessly; during the day it perches most of the time, feeding at night in ponds or in the morning and evening in paddy fields, but is also found in salt water.

Wigeon.

Mareca penelope. *Peasan,* Hindustani.

There seems to be a prevailing idea among people whose knowledge of ducks is limited that any sort smaller than a mallard and yet obviously too big to be a teal, must be a wigeon; but the real bird, which has only two near allies, both American, is quite unmistakable, owing to its small, inch-long, narrow bill, which is blue-grey with a black tip, and its unusually long and narrow wings; the belly is also conspicuously white in both sexes. In weight it is indeed intermediate between the full-sized ducks and the teal, weighing about a pound and a half.

The wigeon drake is very handsome, his chestnut head with yellow forehead contrasting well with his salmon breast and grey back; a large white patch on the wing is very conspicuous in flight. The female's brown plumage is less conspicuously mottled than that of other brown ducks of her sex, but the points above given will distinguish her easily. The male in undress is similar, but much redder brown as a rule; his white and green marking on the wing will distinguish him in this stage.

The sexes differ strikingly in voice as well as in plumage, for the drake utters what Hume well calls "a whistled cry," pre-

1/3
MARECA GIBBERIFIONS

sumably imitated by the Hindustani *Peasan* and the Nepalese *Cheyun;* the female growls, but is far less noisy than her mate. When in flocks, the frequent "whewing" of the drakes is very noticeable, as also the way in which they apparently hump their backs, by raising the ends of the wings and depressing the tail, this being their display; in the ordinary way wigeon float rather high, and are recognizable by their small heads and bills and pointed tails. They fly lightly and fast, wheeling and turning with ease.

They are common winter visitors to India, Burma and Manipur, but do not penetrate further south than Mysore, and in many localities are rather uncertain in their appearance, not turning up at all in some years; in Bengal, from what I observed in the Calcutta Market, they are regular enough in their appearance. Although to a certain extent omnivorous, as all ducks are more or less, they are specially vegetable feeders, and have a particular fondness for grass, so that they feed on land more than most ducks, and especially affect pieces of water with meadow-like turfy margins. They frequent salt water to some extent, but are not so much sea-coast birds as at home, where they are among the chief quarry of the sea-coast gunner, feeding on the sea-grass so common on the British coasts. Their table qualities in India are rather uncertain, and they do not rank so high in this respect as at home.

It is noticeable that they are inclined to avoid a district in abnormally wet years, which at first seems a curious thing for ducks to do; but the effect of too much water is to cut off their food supply to a great extent, the shore grass being drowned, while the water-weeds are too deeply submerged for surface-feeders as they are; though they can dive well if pressed, they do not seek for food in this way as a rule at any rate, though a captive bird in France has been known to do so during a flood.

The wigeon is a high northern bird in its breeding range, but is more of a western than an eastern species at all times, and now and then strays to North America, where, however, the ordinary wigeon is the distinct *M. americana. Patari* and *Pharia* are Hindustani names as well as that given, and in Sind the bird is called *Parow.*

Pink-headed Duck.

* *Rhodonessa caryophyllacea.* *Golabi sir*, Hindustani.

The pink-headed duck stands quite alone in coloration among our birds. Its body is as black as ink—the brownish Indian ink; its head is as pink as new blotting-paper, in the case of the drake at any rate; the duck's head is like the same pink blotting-paper after it has become faded and soiled, with a long black blot on the crown. Her plumage generally is duller and rustier than the drake's, and her bill is black, whereas his is fleshy-white; but the general resemblance is close. The young are also duller than the old drake, have drab heads, and underneath are of a dirty mottled brown; but the whole family are easily distinguished by the buff tint of the quills of the wings, noticeably contrasting in flight with the dark body. Young drakes assume the white bill before getting the full plumage, and the old ones in undress have the duck's black crown-streak, but otherwise do not change colour.

The size of the pink-headed duck is about that of the mallard or spotted-bill, but it is more slenderly built, the head and neck being positively lean, and the latter is generally carried with a backward curve. The weight is about a couple of pounds.

This most extraordinary duck is a resident with us, but unfortunately has a very limited range, being practically confined to certain districts of Upper Bengal, being fairly common in Purneah and Tirhoot, and also found in Bhagalpur and Maldah; outside this district it is rare everywhere, though stray specimens have turned up as far away from its home as Nepal, Delhi, Bhamo, and Madras, which localities about mark off its limits to the north, west, east, and south. Latham, writing a century ago, said it was common in Oudh; but even if his information was correct then, it is as rare there now as in the north-west generally. I fear it is getting rarer still, as when I was in India in the nineties one could generally see about half-a-dozen in the Calcutta Market in a winter, though as much as Rs. 15 each would be asked for them; they were kept alive,

---

* *Anas* on plate.

MARECA ♂ PENELOPE

having a well-known value as ornamental birds; but now, I am told by friends from Calcutta, that an offer of Rs. 100 per bird would probably produce not a single specimen. Yet not many have reached Europe alive so far, and none, so far as I know, have bred in captivity anywhere.

It is just possible, of course, that the birds have not been extensively caught or shot out, but simply "shyed," as our bird-catchers say, by too persistent netting, as they bear so high a value. As I said in my book on Indian ducks, I think this bird should not be killed at all; it would be no loss to the game list, being a bad table bird, and not numerous enough even in its metropolis to be a regular object of pursuit. It is usually seen in small parties only, but flocks of thirty are on record; naturally in the breeding season pairs only are to be seen. Fortunately the localities the birds mostly affect are out-of-the-way bits of standing water, well supplied with reeds and other cover, and situated in forest. The bird, however, unlike most of our resident ducks, is not a percher, and nests in the high grass, sometimes hundreds of yards from water. The eggs are as extraordinary as the bird itself, being so rounded and white that they look almost like unpolished billiard-balls. The number laid varies from five to twice as many, and the nests containing them are round and well formed, the materials being dry grass and a few feathers; the laying time is June and July.

The ordinary habits of this duck appear to be much the same as those of the spotted-bill, that is to say, those of an ordinary surface-feeder; but as the bird would appear to be allied to the pochards, I may mention that I once saw one dive as neatly and as long as a pochard, though why I cannot say. The duck has been seen to employ most elaborate affectations of injury to decoy intruders from her young, and her note was described as loud quacking. I have never heard it, but the drake's is a mellow two-syllabled call, which I have tried to render as "wugh-ah." The wing-whistle is very characteristic, if the flight of the bird in an aviary is any guide; it is clear yet soft, and the flight is easy, and in the open rapid and strong. Of the coloration of the down of the young and of the drake's display nothing is known apparently.

The bird is thoroughly well known to the natives. Besides the Hindustani names of *Golabi sir* or *Golab-lal-sir*, it has the Bengali one of *Saknal*, and that of *Dumrar* or *Umar* in Nepal and Tirhoot.

Common Pochard.

* *Nyroca ferina*. *Burar-nar*, Hindustani.

Being the original and typical pochard, this bird, as Hume says, really does not need a qualifying epithet; "red-headed," that commonly applied to it, is no more distinctive than the Hindustani *Lal-sir*, since both this and the red-crested species have red heads. *Pochard*, by the way, means in French a drunkard, and I rather fancy it was applied to this species on account of the red eyes, which are a most noticeable feature in the drake. We all know the text, "Who hath red eyes and carbuncles? Those that follow after strong drink."

It is true that authors persist in saying the eyes are yellow, but this is really rarely the case in the living male, though the eye may turn yellow after death, or even temporarily during extreme fright, as when the bird is handled. In females, as far as I have seen, the eyes are almost always brown, but I once saw a red-eyed specimen. This pochard is a squat, thick-set, big-headed duck, with a very short tail, which is not noticeable on the water, as the bird swims very low, especially astern. The male's pale grey body, contrasting with the black breast and chestnut head (which at a distance also looks dark), is very characteristic both on the water and on the wing, especially as the latter has no conspicuous mark, being grey like the body. The female is also grey, but dull and muddy, with the head and breast a decided brown. The absence of any white on the wing will distinguish her from all the other diving ducks except the rare stiff-tail, which is sufficiently well characterized by its peculiar beak and tail, to say nothing of other parts. Males in undress have the breast grey instead of black. This is a heavy duck for its size, weighing about a couple of pounds or little under, and showing but little sex difference in weight.

---

* *Fuligula* on plate.

♂ FULIGULA FERINA.

The pochard is as well known a bird in India as in Europe, but only as a winter visitor, ranging throughout the north to Manipur and even Burma, but not going further south than Bellary. It is very common in the northern part of its ranges but comes in rather late, not till the end of October even in the north as a rule.

It goes in flocks numbering at times thousands, but of course commonly in dozens, according to the nature of the accommodation, the big flocks being found on the big sheets of water. It is fond of open water, and may even be found on the sea coast. A few feet of depth at least is in its favour, for it is one of the finest of divers among our Indian ducks, and gets nearly all of its food in this way, though now and then small flocks may be found surface-bibbling in the shallows, and in rare cases feeding on land. Their gait in this case, as Hume says, is certainly very awkward, but in practice they walk rather better if put to it than most diving ducks, in spite of the size of their feet and eminent adaptation for swimming and diving. They rise with considerable trouble and exertion, as one would expect from their small wings, and blunder into standing nets in numbers. Their wing-rustle is said to be characteristic, as would be expected from the quick action necessary in flight to such a full-bodied small-winged duck as this is.

The vocal note is the sound like *kurr, kurr*, which takes the place of a quack in several diving ducks ; the male has a separate note during courtship of a wheezing character.

This pochard being a near relative of the far-famed American canvas-back, it is not surprising to find it a good table bird ; in fact, it is the best of the Indian diving ducks, and the only one to be relied on for quality among these, except on the sea-coast. Its food is vegetable by preference, consisting of water plants, rice, &c., with an addition of water snails, which are eaten by practically all ducks, and other animal *hors d'œuvre.*

This pochard is of a naturally tame disposition, but gets wary enough when persecuted at all ; and winged birds will give plenty of trouble to bring to book, though Hume considered them less elusive than the white-eye. However, as he says, they are generally shot in more open water. As far as actual diving goes I should say the common pochard is really the better performer of the two.

Although a very similar species inhabits North America besides the canvas-back, this really has a yellow eye and seems to me to be just separable after having seen live specimens; so the pochard may be reckoned as purely an Old-World bird. The female is distinguished as Dunbird by English sportsmen, but the same native names seem to apply to both sexes. *Cheun* is used in Nepal and *Thordingnam* in Manipur.

White-eyed Pochard.

* *Nyroca ferruginea.* *Karchiya*, Hindustani.

The little bright-chocolate white-eyed pochard is far the commonest of Indian diving-ducks, and in winter may be looked for anywhere south to Ratnagiri and east to Manipur, but it is more a northern than a southern bird with us, like all the pochards.

The white eye is only a masculine character, but develops early, the female's eyes being dark grey and not conspicuous. Moreover she is not nearly so richly coloured as the drake, being especially clouded with blackish about the head. The coloration, however, is of the same general type, both giving the impression of a small dark duck, showing no white except on the stern when on the water, but displaying a conspicuous amount on the wings when in flight. Young birds are not rich dark brown, but dirty light brown, much like the colour of brown paper, certainly not the gingery orange shown in the background figure in Hume's plate. In all, the upper parts are darker and devoid of red tinge.

This is not only the commonest but the smallest of Indian pochards, seldom weighing much over a pound and a quarter. Except in Kashmir, it is only a cold-weather bird, coming in late in October. When in residence, they prefer before everything water well covered with weeds, or with plenty of rushes; just those places, in fact, which other diving-duck tend to steer

* *Aythya nyroca* on plate.

1/3

AYTHYA NYROCA

clear of. But this species likes cover, and also packs much less than ducks generally, half a dozen being a far more likely party to come across than half a hundred, extremely numerous as the birds are in certain districts. They also often go in pairs or even singly, and in any case have a habit of getting up independently like quail, which of course endears them to the sportsman on account of the number of chances he gets. The flight also resembles a quail's in being low as a rule, and often terminating in a sudden stop ; it is fairly fast, though the bird starts awkwardly, rather like a coot, which in some ways it resembles in habits, just as the goosander does a cormorant; so distinct are the ways of ducks in reality, though people seem to look on them as monotonous uninteresting birds. Winged birds are notorious for their skill in taking cover and disappearing altogether.

It must not be supposed, however, that the white-eye is confined to weedy water; it is a most versatile bird, and will put up with any sort of aquatic habitat at a pinch, so long as it is actually in water, for it does not seem to feed on land. Thus it may be found down on the sea-coast or up in the pools of the hill-streams, on lakes without cover or even on rivers, and, unlike all other diving-ducks, in little stagnant pools. On land, although not quick or elegant in its movements, this species seems less out of its element than other pochards, as I have known it live well in an aviary in the Calcutta Zoo with only a very small tank; but then I have also known even smews do this at home, so perhaps it is only an evidence of constitutional toughness. In feeding, this duck is particularly omnivorous, vegetable and animal food being much alike to it; it is thought but little of for the table, as a rule, but I could never see why it was so much abused; but then I am fond of ducks generally, as well for eating as for observation. The note is a *kurr* in the female, a weak, faint quack in the drake, which jerks back his neck in a curve when courting her.

The white-eye breeds in Kashmir only in our limits, nesting late as a rule, for when it was the custom to take wild ducks' eggs for sale, this poaching traffic did not begin till June. About half a dozen eggs make a sitting, though more occur; they are drab or some shade of buff or pale brown. The nest is made of vegetable matter chiefly, there being but a small amount of down lining supplied by the bird in many cases. It is sometimes actually in the water, supported on the weeds, and in any case very near the edge.

Outside India the white-eye breeds through Western Asia to the Mediterranean region; it is really a southern species, rarely going north of Central Europe; in Britain, it is the rare "ferruginous duck" of the bird-books. India seems its chief winter resort, as in the case of the splendid red-crested pochard. The Bengali name is *Lal-bigri* or *Bhuti-hans; Malac* is used in Nepal Province and *Burnu* in Sind.

Baer's White-eyed Pochard.

Nyroca baeri. *Boro Lalbigra*, Cachari.

That Baer's white-eye is a very uncertain visitor to our Empire, is proved by the fact that Hume and his numerous collectors and correspondents never got to know about it. Yet the adult is a most unmistakable bird, the green-glossed black head contrasting so strikingly with the chocolate breast which it shares with the common white-eyed pochard. In fact, except for the head, the two are much alike, but an important difference is that the white on the belly of the Baer's white-eye runs irregularly up on to the flanks, thus showing above the water-line, and furnishing a means of distinction, even at a distance, from the common white-eye.

The sexes are more alike in this species than in the ordinary white-eye, but the female is distinguishable by the presence of a rusty area between the eye and the beak, contrasting with the green of the rest of the head. This, like the plumage generally, is less rich than in the drake, but the difference is very slight, and it is quite a mistake to describe the female's head as simply black. When a bird shows no green gloss it is generally small, and probably indicates a cross with the commoner species, which is very noticeably the smaller of the two.

So much is this the case, that it is one of the distinctions of the young birds, which in this species are dull light-brown as in the last, but have a rusty tinge about the face and a distinct black shade on the crown which is not found in young common white-eyes. The difference is especially marked in the bill, which is about half an inch longer in the eastern than in the western white-eye; in fact, the whole bird is longer and less dumpy, though the family resemblance is most obvious and close.

Even in its ordinary wintering-places in China, the eastern white-eye seems somewhat irregular in its occurrence, and little is really known about it except that it breeds in East Siberia. There has certainly been a considerable winter westward movement of the species of late years, beginning apparently with the year 1896, when it turned up in the Calcutta Market, by no means an unexploited locality. The rush appeared to culminate in the next winter, the birds then becoming gradually scarcer; in 1902, up to December when I left India for good, there had been none in; but about February was about the likeliest date for them; in 1896-7 they were as common as ordinary white-eyes. Mr. Baker also got them, after the occurrence of the species here was made known, from Cachar, Sylhet, near Bhamo, and the Shan States, which is what one would expect, although the birds do not seem to have been numerous, as they apparently were in Bengal; he only got three from Burma, for instance.

Although not recorded on the Continent, the birds even pushed as far as England, where two have been shot of late years, one at Tring, and one while this book was being written, in Notts. The only observation worth recording here I was able to make on these birds, of which I kept several alive in the museum tank, and got others for the Calcutta and London Zoos, was that when kept full-winged in an aviary they rose as easily as surface-feeding ducks; this may mean they escape netting much more than other pochards. I may also mention that the note and courting gestures of the species are, as one would expect, like those of its common ally, and it is certainly no better to eat, according to those who have tried it. As confirming the view of those who attribute the abnormal lingering of pairs of migratory birds in India to some injury incapacitating one partner from migration to the northern breeding-grounds, I may, in conclusion, cite the case of an unpinioned male of this species I had, which remained in the museum tank for at least two summers along with a pinioned pair; indeed, I never even saw him fly, and ultimately I caught him when he was in moult and gave him away to go to Europe along with the pinioned birds. Of course he and not the pinioned drake might have been mated to the female, but even if he were not, he evidently did not like to leave his companions, and his constancy rather tends to show that flocks of the species never passed over during his stay with me, or he might have been tempted to do so.

Tufted Pochard.

* *Nyroca fuligula.* *Ablac,* Hindustani.

The tufted pochard drake is conspicuous among all our water-fowl, not so much by his long thin drooping crest, which is only noticeable close at hand, but by his striking magpie coloration, black in front and behind and white in the middle; the back, indeed, is black as well as the breast, but the broad white flanks are what catch the eye as the bird swims. The female is dark brown with a much shorter crest, and seldom shows any white above the water-line.

On the wing both sexes show as small dark ducks with nearly white wings, and at a distance may be easily confused with the white-eye, but differ somewhat in habits and choice of location. Young birds of the year are dull light brown, very like young white-eyes, but may be distinguished, if the crest and the characteristic yellow eye of this species has not developed, by the much shorter and broader beak, especially wide at the tip.

Although a small duck, the tufted is bigger than the white-eyed pochard, being broader built and averaging about half a pound heavier. In spite of this, however, it is the most active flyer of all the common pochards, getting sharply off the water

* *Fuligula cristata* on plate.

FULIGULA ½ CRISTATA

and flying fast and without undue effort. Its ways may be studied to perfection within a hundred yards or so of the India Office, for in St. James's Park it lives and breeds in complete liberty, and specimens visit all the London waters.

In India it is a winter visitor only, and extends east to Manipur and Burma, but not further south in India than North Coimbatore. It was very commonly brought into the Calcutta Bazaar in my time, but could not compare in numbers with the white-eyed pochard, which was by far the commonest diving-duck to be met with there, and commoner than any of the migrant ducks except garganey and common teal. I may here observe that the light brown colour here attributed to young tufted pochards may indicate a special Eastern race, or more probably be due to fading under bright sunlight, for those hatched in England fledge off as dark as the old female. I have, however, seen two young English drakes which had assumed as their first plumage the undress plumage of the old drake, which in this stage has the white flanks obscured by a plentiful pencilling of fine black lines, so that they look a smoky grey.

In India the tufted pochard is seen in both small and very large flocks; it likes broad open water, and is very difficult to shift from its chosen location on the first day of shooting, though after a day's worrying it will not come back to be shot at again like many other ducks! It is a fast swimmer, and as a diver at least equals the common red-headed pochard; its evolutions under water may be watched with profit in St. James's Park, where it may be seen to go over a lot of ground before rising to the top. Sometimes a flock will on Indian waters prefer to dive rather than fly, in which case, if one's boat can be pushed on to where they went down, they will afford a good many snap-shots—at the risk of casualties from shot glancing off the water! When in pairs, the duck rises first, calling *kerr* as she goes.

Although it greedily devours bread in semi-domestication, it feeds mostly on animal food—water-snails, small fish, and so forth—in the wild state, and appears never to feed on shore; even when petted in a park it seldom leaves the water for bread, and its gait on land is particularly hobbling and awkward, much worse than that of the common pochard. It is a day-feeder, and in spite of its animal-feeding habits is often good eating, though too frequently what Hume calls "froggy." It comes into India early in October and may remain as late as May. It is a widely ranging bird, extending to Norway in the breeding season, and to North Africa in one direction and the Malay Peninsula in the other, in winter. In Hindustani, besides the name given above, which expresses the male's pied colouring, it is called *Dubaru* and *Rohwara*, while in Nepal it shares the name of *Malac* with the white-eye; in Sind it is called *Turando; Nalla chilluwa* in Telugu.

Scaup.

**Nyroca marila.*

The scaup, which in winter-time at any rate is chiefly a sea-bird, feeding on shell-fish, has rarely occurred in India, and when it does turn up in the white-faced brown immature dress is likely to be confused with the young of some of the other pochards, to which group it belongs, its nearest ally here being the tufted pochard, which is, indeed, called by some writers tufted scaup.

The old drake and duck, however, are easily recognizable close at hand, the former having a deep green-black head, black breast, pencilled-grey back and white flanks; the latter having a brown head and breast, but also a grey back, and a very distinct white face. The female tufted pochard often shows some white here, but always has a dark back. Like tufted pochard, scaup have yellow eyes and broad bills, but they are considerably bigger, about equalling common red-headed pochards in size. From these they can easily be distinguished by the white, which, like tufted and white-eyed pochards, they have on the wings.

The few scaup which have been reported from India have turned up in widely separated localities, from Kashmir to Lakhimpur, and south to Bombay and Chittagong. In Oudh it would appear to occur fairly frequently, the Rev. J. Gompertz having shot eleven between 1897 and 1904 inclusive, as recorded by Captain Wall, quoted by Mr. E. C. S. Baker, and other records have before been made from the same province.

---

* *Fuligula* on plate.

FULIGULA ♂ MARILA

The scaup is a bird of the high north, but in winter is found as far south as the Mediterranean, South China, and Guatemala, for it inhabits America as well as the Old World. It is not difficult to get near, but is extraordinarily tenacious of life and a most energetic and rapid diver when wounded; while when at length captured it is not good eating, so that its rarity here is not a matter for much regret.

Red-crested Pochard.

* *Netta rufina.* *Lal-chonch*, Hindustani.

The big red bill and bushy chestnut head of the big red-crested pochard distinguish him from afar, to say nothing of his strongly contrasted body-colouring, black at breast and stern, and brown and white amidships; the black runs all down the under-surface, though on the water only the white flanks are seen.

The female, although with no bright colours, being merely brown, with cheeks and under-parts dirty white, is easily distinguished from other ducks; she also has a bushy-looking head and in particular a black bill tipped with red.

The drake in undress plumage hardly differs from her but in having more red on the bill—in fact, in some specimens this remains as completely red as in the full plumage.

The red-crested pochard is the biggest of all our diving ducks except the goosander, drakes weighing about two and a half pounds and ducks only about half a pound less.

The flight is less heavy and whizzing than that of pochards in general, but the wing-rustle is usually distinguishable from that of the common pochard, being louder and harsher.

These birds are very abundant in many places in India, flocks even of thousands occurring, which look, from the bright-coloured heads of the drakes, like beds of aquatic flowers; they come in in October and November and leave about April. The big flocks tend to split up into parties of a few dozen where there are not very large pieces of water, but stray specimens may turn up

---

* *Branta* on plate.

almost anywhere. What they chiefly like, however, are expanses of deep, still water with plenty of weed on the bottom, and so they are chiefly birds of broad sluggish reaches of rivers and extensive lakes and jheels. They are found all through Northern India east to Manipur, but do not commonly go south of the Central Provinces, though said even to reach Ceylon occasionally. They swim fast and dive well, getting much, if not most of their food in this way during their stay in India; but they also frequently feed in shallow water by turning end up, and generally behaving like surface-feeding ducks, even coming ashore to feed; and it is a curious thing that in captivity, even on a large piece of water, they very seldom dive, although other pochards constantly do so. They are less clumsy in shape than these, and do not walk so awkwardly.

The food is very varied, but more vegetable than animal; water-plants, grass, insects, frogs, and even small fish, all enter into their *menu*. They are generally fat, and are sometimes as good as any duck, but may also have what Hume calls a "rank, marshy, froggy flavour." They are among the best sporting birds in India, being wary and shy; in fact, Hume considered them the hardest to get at of all the ordinary quarry of the wild fowler in the East.

They are generally day-feeders, but also feed at night, and are commonly shot at flighting time, though then only in small parties. The flocks usually contain both sexes, but occasionally males only may constitute a flock. The sexes differ considerably in their voice, the drake's note being a whistle—not the same, however, as the wigeon drake's; the duck's call is the usual *kurr* of the females of the pochard group.

The great distinctness of the sexes causes this bird to be one of the few with different sex-names in the vernaculars; in Bengali the male is *Hero*, the female *Chobra-hans;* in Nepalese the words are different, *Dumar* for the male and *Sanwa* for the female; the Sindhi *Ratoha* applies to both sexes.

The red-crested pochard is nowhere a duck of the high north; it breeds as near us as Turkestan, and extends west through Southern Europe to North Africa. India seems to be its chief winter resort. In Britain it is rare as a wild bird, but well

BRANTA ⅓ RUFINA

known in captivity, in which state it often breeds. The nest is on the ground in rushes, and the eggs, when fresh, are remarkable for the brightness of their green colour, about eight being the usual clutch.

Stiff-tailed Duck.

Erismatura leucocephala.

The remarkable appearance of this duck always attracts attention; it sits very low in the water, often erecting its long thin wiry tail, which balances as it were the big head with its remarkably broad bill, much bulged at the root. When approached it dives in preference to flying, and if it does rise does not usually travel far; the wings are extraordinarily small. The plumage is peculiar but not striking, being of a pencilled brown, sometimes much tinged with chestnut. The head is marked with a lateral streak of white on a blackish ground, and the throat is white, the bill and feet slate-colour. At least, that is the plumage of most of the specimens which turn up in India; the adult male is a really striking-looking bird, with a sky-blue bill and snow-white head and throat, set off by a black crown and neck. Although extremely broad, this bird is hardly longer than a common teal, and its wings are considerably smaller than in that species; yet the weight must be twice as much, and it is a wonder how the bird manages to travel at all. The wings fold up very closely as in a grebe, and are of a plain drab colour without any mark. The tail is not always of the same length, but its wiry character and the scantiness of the coverts at its base are distinctive.

The stiff-tail is a duck of rather unsocial habits and never seen in large flocks; those found in India have generally been alone or at most in pairs. They are found on rivers as well as in pools, and are probably pretty widely distributed, specimens having occurred from Kashmir to the Calcutta Bazaar. Here I once got a live one, but this unfortunately had one leg hopelessly disabled, and moreover would not or could not eat, so I was reluctantly obliged to make a specimen of it. In spite of its affliction it was so tame that it would plume itself while being held in the hand. Grebes will also do this, so that even in captivity this species retains the grebe-like ways which characterize it when wild. It may also be mentioned that the under plumage, though buff and not white, has the silvery lustre of a grebe's. Very grebe-like was the behaviour of a female procured at Peshawar by Captain Macnab, I.M.S., in 1899. A hawk also tried to "collect" it, but as soon as he made his point above, the duck went under, and after coming up close by dived again, till after about five minutes the hawk went off in disgust. The tail, at full cock when swimming, was straightened out as the bird went down. The call is said to be a grating, quacking note, and the food to consist of small water creatures and vegetable substances. But, as a matter of fact, little has been observed about this bird's habits, though it is widely distributed in a sort of central zone, from the Mediterranean region east to our borders. Personally, I believe it breeds in India, because a specimen shot near Hardoi was moulting and had no quills grown. None of our winter water-fowl moult while they are with us, so far as I am aware, while the residents can moult at any time they like, having no long journey to take. However, one must not forget that it is in its winter quarters in South Africa that the common swallow moults.

The nest in any case is nothing out of the ordinary, being built among the waterside vegetation and composed of it; but the eggs are, being remarkable for their coarseness of surface and large size in proportion to the bird, though their colour is simply white. The ducklings are dark brown in the down, with white on the under-parts, conspicuous on the throat and sides of neck, and there are some faint white spots on the upper surface.

Golden-eye.

Clangula glaucion. *Burgee,* Punjabi.

It is not surprising that this, the genuine golden-eye, should be often mixed up by Indian sportsmen with the tufted pochard, since both are pied diving-ducks with yellow eyes; but in reality they are quite easily distinguishable even at a distance, for in the

1/3
CLANGULA GLAUCION

real golden-eyed drake the breast is white as well as the flanks, and there is a great white patch on the face. Even the female, whose breast is grey like her back, has a white neck sharply contrasted with her dark-brown head, and so is well distinguished from the female tufted duck, whose neck is dark continuously with the head and breast.

In flight the golden-eye is distinguishable by sound more than any other duck, the very loud and clear whistle made by the wings, which have no white on the end-quills like those of the tufted pochard, being a most marked characteristic of the species, and often giving it a special name, such as "rattle-wing" and "whistler."

There is a good deal of difference in size between the sexes of this species, the drake weighing two pounds or more, sometimes nearly three, while the duck runs more than half a pound less, and looks conspicuously smaller when both are together. Young males are very like females, but the old male in undress may be distinguished, not only by his much greater size, but by having more white on the wing.

The golden-eye has a bushy-looking head and a very narrow, short, high-based bill ; the tail is also very characteristic, being longer than ducks' tails usually are. Often, however, it is not to be seen as the birds are swimming, being allowed to trail in the water, but sometimes they float with tail well out of the water, when the length becomes noticeable. On land, where they spend but little time, they stand more erect than most ducks.

Golden-eyes in India only appear as uncommon winter visitors, as a rule ; but in the valley of the Indus at one end of our area and the Irrawaddy at the other, they seem to occur regularly, and are also common in the Lakhimpur district, frequenting hill-streams like the mergansers, to which, rather than to the ordinary diving-ducks, they are related. Indeed, this species has been known to produce hybrids with the smew in the wild state, and, like that bird, it breeds in holes in trees in the northern forests.

The golden-eye, however, is found all round the world, not confined to the Eastern Hemisphere. It goes in winter either singly or in flocks, and Mr. Baker has shot one consorting with gadwall, and noticed that it flew well with them; in fact, this is the most active flyer of all the diving-ducks except perhaps the smew. In the water it is a fine performer, and catches fish like the mergansers, also feeding on shell-fish and water-weeds. In diving it seems to slip under, as it were, more neatly than the pochards, the fanning-out of the tail being conspicuous as it disappears; and it has been noticed that when a flock are below and an alarm comes from above, all the diving individuals will rise and make off out of shot, not coming up and giving the enemy a chance as the pochards will do. The general character of the bird is, in fact, one of extreme wariness, and in this respect it is a merganser rather than a true or typical duck. The alarm-note is given by Mr. Baker as "a loud squawk," but this is no doubt uttered by the female only; the male's note is different, as is the case with the male mergansers, which this bird resembles in having a large angular bulb in the windpipe. A peculiarity of its beak, unique among Indian birds, should be noticed; the nostrils are very near the tip, further forward even than in a goose. As an article of food this bird is highly fishy; but its green eggs are esteemed by the inhabitants of the north, who put up boxes for it to lay in.

Smew.

* *Mergus albellus.* *Nihenne*, Sindi.

The pretty little smew is at once known from all our diving-ducks by its short narrow dark beak; its quick nimble way of rising into the air is also distinctive among this splattering tribe, though it often prefers to swim, which it can do at great speed, rather than fly. It can swim either high or low in the water, and is exceedingly wary.

The adult male's plumage is very distinctive and beautiful, being white below and on the head and neck, except for a black patch on each side of the face, like a mask, and a black V at the back of the head, which has a full, though short crest.

* *Mergellus* on plate.

MERGELLUS ALBELLUS.

The rest of the plumage is black, white, and grey, no other colour appearing. The weight is about a pound and a half. Females and young males, which are much smaller, the male not attaining either full size or colour till his second year, and much more likely to be met with, are dark grey above and white below, with bright chestnut head and white throat, and black-and-white wings. They weigh little over the pound. The short narrow beak and bright chestnut head with pure white throat extending well up on the jaws make the name of "weasel-coot," or its equivalent "vare-wigeon," sometimes given as this bird's old English names, quite intelligible, as there is something decidedly weaselly about the bird's look. In its extreme activity also this little fishing duck recalls the smallest of the four-footed carnivora; it is the fastest diver of all our waterfowl, flies with ease and speed, and even on land, in spite of the breadth of its body and shortness of its legs, to say nothing of its very large feet, moves quite quickly and looks far less ungainly than many ducks one would expect to walk better.

In this activity this bird resembles the mandarin duck, which also, by dint of sheer energy of movement, is able to hold its own with other ducks in departments for which they seem to be better adapted structurally. Another coincidence between these two pretty species is the fact that they both breed in holes in trees; but the smew is even less likely than the mandarin to be found breeding here. So far it is only known as a winter visitor to the north of India, from Sind to Assam; it does not appear to go further south than Cuttack, and though fairly well known in the North-west is not abundant anywhere. It is generally seen in flocks of about a dozen, but pairs or single birds may occur. In the case of one such which was shot, the flesh was found to be quite good eating, which was rather unexpected, the smew generally having a particularly bad name for excessive fishiness of flavour. Besides fish, however, it feeds on water-insects and shellfish, and even aquatic plants.

It is a widely distributed bird, ranging all across the Old World, though it only breeds in high latitudes. The courtship is very interesting to watch, the bird swimming about with the head drawn back proudly like a miniature swan, and the fore-part of the crest raised, while now and then he rears up in the water with down-bent head.

Goosander.

* *Merganser castor.*

The goosander is a fishing duck built on the lines of a cormorant; narrow head, long flat body, with legs far astern, rather long tail, and especially long, narrow, hooked beak; in fact, many people on the first sight of one hardly realize it is a duck at all. However, its striking variegated plumage is quite different from the crow-like coloration of the cormorants: the drake is pied, being below white, with the head, upper back and part of the wings black, the lower back and tail grey, and bright red bill and feet. The head has a green gloss, and the under-parts often show a wash of salmon or apricot colour.

The female is very different, being bright chestnut on the head, which in her is well crested; French grey above, about heron or pigeon-colour, and white below; the wings are black and white and the legs and bill red, much as in the male, but not so distinct in colour. The male in undress much resembles her. The beak, it will be noticed on close inspection, is set with backward-pointing horny teeth, and has not the wide gape of a cormorant's, and the feet are quite ordinary duck's feet, having the hind-toe short; not large and webbed to the rest like a cormorant's, so useful both as extension of paddle and a perch-grip. The goosander is one of our largest ducks, the male weighing about three pounds, while even four and a half has been recorded; females are generally less. They are usually very fat, and, according to Hume, will make a good meal if skinned, soaked, and stewed with onions and Worcester sauce, and if one has not got any other form of meat or game available. This is quite likely to happen where goosanders are shot, as their haunts are different from those of ducks in general; they are birds of the hill-streams chiefly, being resident in the Himalayas and

* *Mergus* on plate.

½
MERGUS CASTOR

merely moving up and down according to season ; in winter they may be found all along the foot of the Himalayas, in northern Burma, the south Assamese hills, and even as far south as the Godavari and Bombay, where E. H. Aitken once shot one on salt water.

Their food is mainly fish, though they eat other live things as well, and in captivity will feed on raw rice freed from the husks ; they are extremely greedy, and will eat over a quarter of a pound of fish at a meal, digesting bones and all. They are therefore not birds to be encouraged where the fishing is valued, and there is likewise this excuse for shooting them, that they are likely to be appreciated by natives who, like our Elizabethan ancestors, like a good strong-tasting bird.

Moreover, they are really sporting birds ; they are wary and require careful stalking, and when hit are by no means booked, as they will dive literally to the death. Mr. Baker records a case in which a female, after being hit, managed to keep out of range of his boat, propelled by two men, for half an hour, and then appeared on the surface dead, having died while diving, game to the last. They are naturally fast swimmers as well as good divers, and though they often, when floating quietly, sit nearly as high out of the water as an ordinary duck, also swim low with the tail awash, and when wounded or frightened show only the head and neck above water, much like a cormorant. They also have the cormorant-like habit of sitting erect on the shore, partially expanding their wings, though here again their carriage is often level and like that of an ordinary duck. So it is when walking, when they look less awkward than some diving ducks ; when running on land, and this they can do well at a pinch, they stand very erect.

They resemble cormorants, too, in often fishing in concert, forming a line across the stream and all diving together, so as to drive the fish before them. Although perhaps preferring the stiller reaches and pools, they are at home in the most rapid and rushing torrents. They are slow in getting on the wing, but fly fast when well up. The note of the drake is a croak like "*karr*"; of the female a distinct quack.

Although the young, which in the down are brown above,

tinged reddish about the head and neck, and marked with a few white patches, and white below, have been taken in the hills, the eggs have not yet been found in India. They are cream-coloured and very smooth, and number over half a dozen. The nest, well-lined with down, is generally in a hole in a tree or bank in the birds' known breeding-places, which extend all round the world in the northern regions, the supposed distinctness of the American race resting on the most trivial characters. The female often carries her young on her back when swimming. It is a curious thing that there seems to be no native name recorded for this most conspicuous species.

Red-breasted Merganser.

Merganser serrator.

The red-breasted merganser, which is, like the goosander, found all round the world in the Northern Hemisphere, is chiefly a salt-water bird when it leaves its breeding-haunts on the fresh waters of the north, so that it is not surprising that it should be one of the rarest of our water-fowl. It is probable, however, that it is often confused with the goosander, as the general appearance of the two is very much alike.

The male of the present bird can, however, easily be distinguished by having the under-parts not uniformly white up to the green-black head, but the white neck cut off from the abdomen by the reddish-brown, black-streaked breast; on each side of this there is also a patch of black-and-white feathers, and the flanks are grey in appearance above, being finely pencilled with black. Thus the bird looks darker altogether on the water, and seen on the wing the coloured breast-patch should attract attention. The bird also has a long hairy-looking crest, not a short bushy mane as in the male goosander.

The female is much more difficult to distinguish from that of the goosander, but the distinctions are clear enough if carefully looked into. The crest in the present bird is short and not noticeable, the head itself is dull brown, with hardly a tinge of chestnut, the back is mottled drab, not a distinct uniform grey, and the white wing-patch found in both species is in this one interrupted by a bar of black. The male in undress is very like the female, but has a black upper back.

This species is decidedly smaller than the goosander, but has quite as long a beak, which is, however, much slenderer, less hooked, and shows more teeth. It resembles the goosander in being a greedy devourer of fish, is a fine diver and fairly good walker, and is excessively wary, at any rate in Europe. In India it has only been got once or twice at Karachi, once in the Calcutta Bazaar and once in the Quetta district.

Comb-Duck.

Sarcidiornis melanonotus. *Nukta*, Hindustani.

The comb-duck is often called, even by Europeans, by its best known native name of *Nukta* or *Nukwa*, and the practice is one to be commended, as in all cases where a bird is of a type of its own and unfamiliar to Europeans. Although to some extent intermediate between ducks and geese, the nukta would never have been called a black-backed "goose" were the male no bigger than the female, since she is obviously a duck; he, however, is quite as big as an ordinary wild goose, weighing between five and six pounds, while the female is only about three.

In plumage, however, they are much alike, only the female is far less richly glossed with purple and green on the black upper parts, has the sides dirty drab instead of pure delicate grey, and never displays the yellow patch under the sides of tail, and the yellow streak along the head, which the male has when in the height of breeding condition. At this time his black comb is a couple of inches high, but shrinks down to less than half in the off-season; the female never has one, nor the young male, till he gets his full colour. In immature plumage the birds are brown, not black, but at all stages the combination of white belly with dark under-surface of wings is distinctive of the nukta among our large ducks.

The amount of black speckling on the white head varies a great deal individually; the whitest-headed bird I ever saw was a young male, and he had black on the flanks instead of grey, and was thus like the male South American nukta, which seems to me, therefore, hardly distinct from ours. The African one is now admitted to be the same as the Indian; no other species is known.

The comb-duck is generally distributed over India, Burma, and Ceylon in suitable localities, such localities being open land provided with plenty of reedy marshes, and scattered large trees; treeless country the bird dislikes, as it is a perching duck and roosts and breeds in the trees; nor does it care, on the other hand, about actual forest. It seldom frequents rivers, but may be found on lakes, and in some localities even on small ponds. It will thus be seen that its choice of localities is very different from that of the geese, while it is not sociable like them, being very rarely found in flocks of more than a dozen or so, and commonly in pairs. It associates with no other duck but the ruddy sheldrake, and that not often, as the two birds affect different places; and unless it happens to be in such company, is nôt so wary as one would expect a large water-fowl to be. Most of its time is passed in the water, though it walks as well on land as a goose, and although it feeds freely on rice and land and water herbage, it also partakes of water-snails, insects, &c., like a typical duck. The brown young birds are good eating, but the adults, though not ill-flavoured, are inclined to be hard; they should be cooked and served like geese. On the water the bird sits high with the stern raised, like a goose, but both there and on land the neck is carried in a graceful curve, and when courting the male arches his neck and bends down his head, slightly expanding his wings after the fashion of a swan, only much less. The comb-duck swims well, and dives vigorously if pressed; in flight it is intermediate in style between a duck and a goose; the male, conspicuous by his size and comb, acts as the leader. It flies and feeds by day, retiring to the trees at night; it is usually very silent, but the note when heard is variously described, sometimes as loud and goose-like, sometimes as a low guttural quack, or, in the case of the male, as a grating sound; probably only the female has the loud call.

The pair seem much attached, and the male accompanies

¼

SARKIDIORNIS MELANOTUS

the female in her search of a nesting-site in the trees; such a site is a hole, or the place where several large branches diverge; an old nest of another large bird has been used, and even a hole in a bank; and the nest has even been said to be sometimes placed on the ground among rushes by the water. The eggs are of an unusually polished appearance for a duck's, and yellowish-white; about a dozen are laid, some time between June and September. The ducklings in down are brown and white above and white below.

The nukta is as well off for names as might be expected; in Telugu it is *Jutu chilluwa*, in Canarese *Dod sarle haki*, and in Uriya *Nakihansa; Neerkoli* is the name in Coimbatore, and *Tau-bai* in Burma, though the Karens call it *Bowkbang*.

Cotton-Teal.

Nettopus coromandelianus. *Girri*, Hindustani.

The jolly little cotton-teal, smallest of Indian ducks, is not a teal properly speaking, and is indeed sometimes considered to be a kind of goose; this, however, is also wide of the mark, and the bird and its few relatives really stand very much alone, their nearest ally probably being the nukta.

In fact, the male's coloration is very much that of the nukta drake in miniature, the lustrous green of the upper-parts and wings contrasting with the general white hue of the head and under-parts and the grey flanks: but the broad black necklace is very distinctive, as is also the white patch on the pinion-quills, only noticeable in flight, but then very conspicuous. The female is brown above and shades into white below; there is a dark eye-streak as well as a dark cap, and the neck has dark specklings running into cross pencilling below; she is also like a miniature nukta in colour, but the resemblance in this case is to the *immature* plumage of the big bird. On the water she looks all brown, and is not conspicuous; the male, among leaves, may also be very unobtrusive, his green back and white head giving the impression of water-lilies with white flowers—this is not mere theory, for I have made this mistake myself, having at first taken the heads of the drakes of a flock of cotton-teal for flowers, and not noticed the females at all. Young males are like females; old birds in undress differ from them by retaining the green-and-white wings.

The beak of the cotton-teal is very short and goose-like, but the tail is long for a duck's, and the bird, when nervous, frequently wags it with a quick quivering action. The legs are short and the feet large, and the birds swim and dive well, often diving on alarm; they do not, I think, regularly dive for food, judging from their hesitation when they do so.

Cotton-teal are found over the Empire generally in well watered and wooded districts; they are naturally therefore not to be found in the dry parts of the North-west.

The district where the species is most numerous is Bengal, where it is called *Ghangarial* or *Ghangani*, but it penetrates even to the Andamans and is well known in Ceylon.

It likes weedy places, and small rather than large pieces of water, and may be found even on wayside ditches, and bush-surrounded pits; it is generally seen in pairs or small parties of less than a dozen, though Mr. E. C. S. Baker has seen as many as a hundred in a flock. Its flight is very fast, and at the same time it is an adept at twisting and dodging; Hume never saw it taken by the great foe of water-fowl, the peregrine falcon, the tiny duck side-skidding from the stoop most dexterously, and being below water before the enemy had recovered itself. The flight is generally low, but when thoroughly frightened the birds will go higher.

The only weak point of the cotton-teal, in fact, is its walking powers; it is very seldom seen on land, and when it tries to go fast or to turn round is apt to fall down; but it is not correct to say it cannot walk at all, as when not hurried it moves on land like other ducks, though slowly and clumsily. This leg-weakness is curious, as it is a perching-bird, roosting and building in trees, so that one would expect it to be at least as strong in the legs as other water-fowl, the perchers being usually good walkers also.

The food of the cotton-teal is mainly vegetable; it seems to feed almost entirely on the surface, and pecks rather than bibbles in the usual duck fashion; it does not stand on its head and investigate the bottom like other ducks. As food it is no

W. Foster.

F. Waller Chromo-lith. 16 Hatton Garden, London.

½

NETTAPUS COROMANDELICUS

better than a common house-pigeon, and as it is so very small, only weighing about ten ounces, and is very tame in many places, it is commonly thought hardly worth shooting, besides which it is such a nice little bird that shooting it is rather like firing at a robin or a squirrel; it does not seem right to make game of it. The breeding season does not begin before the end of June, and lasts till August; the birds moult after this, and the drake has his undress plumage in the winter, unlike most other ducks. Holes in buildings as well as holes in trees may be utilized for nesting, and there is reason to believe that the parents, at any rate the female, carry down the little brown and white ducklings like the whistler. The eggs are like miniature nukta's eggs, remarkable for their smoothness and yellowish-white colour; ten is the usual number of the sitting.

The note of the male is one of the noteworthy peculiarities of this pretty little creature; he often calls on the wing, his peculiar cackle being imitated by several native names, such as *Lerriget-perriget* or *Merom-derebet* among the Kols, and the Burmese *Kalagat.* The Uriya name is *Dandana,* and *Gurgurra* is used in Hindustani as well as *Girri, Girria,* or *Gurja.*

East of India this bird ranges to Celebes and China and reappears as a slightly larger but otherwise indistinguishable race, as far off as Australia.

Mandarin Duck.

Æx galericulata.

Except for one specimen shot by Mr. A. Stevens on the Dibru River in Assam, and recorded by Mr. E. C. S. Baker in his book on Indian Ducks, no Indian-killed example of this beautiful East Asiatic duck is on record, though there is evidence that others have been seen, and even in one case shot. These were all females like the one preserved, or males either in young or undress plumage, and therefore in plumage closely resembling that of the female; and in this species the resemblance is extraordinarily close in such specimens.

Thus the mandarin in India has so far appeared as a small brown duck, rather less in size than a wigeon, with a long tail for a duck, pointed wings with pinion-quills edged with silver-grey and tipped with steel-blue, and a very small beak and large eyes. The upper-parts have no markings of any sort, and the abdomen is pure white, but the breast and sides are mottled with brown and buff. The head is greyish and crested, the male having more grey tint and less crest than the female, which also has a narrow white ring round the eyes. The male's feet are orange, the female's olive.

In full plumage the drake is well known to everyone who takes any interest in waterfowl ; the orange-chestnut fans in his wings are unique, and one of these feathers would be enough as a record of the species ; he also has an orange ruff of hackles, and an enormous crest of copper, green, and white, besides showing many other sharply contrasted colours in his plumage, and possessing a bill of the brightest pink-red or cerise. This duck is well known in captivity in India as well as in Europe, being exported from China. Here it breeds, as also in Japan and Amoorland; it nests in holes in trees, and spends much time in them, being a thorough wood-duck and a regular percher. The notes of the sexes are quaint; the drake snorts and the duck sneezes !

The mandarin is quick and active in its movements in walking, swimming, and flying ; and, although feeding much on land, where it often grazes like a wigeon, or searches for acorns in woods, it nevertheless dives for food occasionally, and is more active under water than almost any surface duck. It is poor eating, and not a bird to shoot more specimens of than can be avoided, owing to its beauty and interesting habits.

It only goes in small flocks and does not seem to be an object of sport anywhere.

Ruddy Sheldrake.

Casarca rutila. *Chakwa,* Hindustani.

This showy spoil-sport, so conspicuous in its foxy-red plumage on land or water, and, if anything, more striking in flight, with its broad slowly-beating black-and-white wings making it tri-coloured, is a bird that cannot be overlooked where

¼

CASARCA RUTILA

it is found, and it occurs all over our Indian Empire except in the extreme south of India and Ceylon, where it is rare, Tenasserim, and the islands of the Bay of Bengal. It is a winter visitor in the plains, but breeds in the Himalayas.

Even if it were not so conspicuous by its colouring—and one gets the full benefit of this by its habit of frequenting the most open places—its voice would make its presence known everywhere, especially as it is seldom silent for long, and even when conversing with its beloved mate and unalarmed, has no idea of lowering its trumpet tones, which have something very stirring and picturesque about them.

There is no noticeable difference in the trumpeting call of the sexes, and their colour also looks alike at a little distance; but on close inspection it will be seen that the female has a white face, contrasting with the buff of the rest of the head, which is in both sexes much lighter than the body as a rule. The male also has in some cases a black collar round the neck, which is supposed to be assumed in summer and lost in winter, though in captive birds, at any rate, and probably often in wild ones, the reverse may be the case. Many birds of this species in India are very washed-out in colour, no doubt owing to bleaching, since in England, where the bird is a familiar favourite on ornamental waters, they are always of the beautiful auburn or chestnut tint.

The Brahminy duck, to give this species the name by which it is usually known in India, is a lover of sandy shores and clear open water, and prefers the banks of rivers to any other haunt, being usually seen in pairs. It keeps more on the land than in the water, walking with an upright carriage and very gracefully; when it does swim it is with the stern high like a goose, and its diving powers are rather limited. It seems to be chiefly an animal feeder in India, devouring small shell-fish and other forms of animal life to be found along the water's edge; it has even the reputation, apparently justified in some cases, of eating carrion; but it admittedly feeds on grain, grass and young corn as well even in India, and in our London parks seems to graze nearly as much as a goose, though there it spends an abnormal amount of its time in the water, no doubt because being pinioned it cannot fly about.

It is not good eating, though it is rendered more tolerable by being skinned, as is the case with so many rank birds of this family; and might be very well left alone by sportsmen if it would only let them alone. This, however, it will not do; it has a very practical working knowledge of the range of a gun, and gets up just out of shot, trumpeting out a duet with its partner, which naturally puts all the other fowl on the alert. As a remedy for this, Hume recommends shooting a few with the rifle, which so frightens the survivors as to make them keep their distance to some purpose—so far off will they then get up that other fowl do not consider there is anything to worry about, and disregard them.

This warning propensity is evidently due to natural noisiness and not to public-spiritedness, for the birds are most unsociable by nature, and, although flocks may sometimes be seen with us in winter, in the breeding season the pairs keep strictly separate, and persecute all other water-fowl, of their own species or any other, including even geese. Even in winter, students of the London park water-fowl may notice that the other birds are nervous of them, and even the mandarin, with all his pluck and bounce, shows by his manner that he knows he is taking risks in snatching the bread from the mouth of the ruddy sheldrake.

In Indian limits this bird has only been found breeding at a high elevation in the Himalayas, 10,000 feet and upwards; the nests are in holes in cliffs, and several are found in the same quarter. The eggs are eight in number as a rule, and creamy-white; the ducklings mostly sooty-black above and white below; they will dive for food while in the down, although their parents are strictly surface-feeders. The ruddy sheldrake also breeds from Central Asia west, all along the Mediterranean, and visits China as well as India in winter. In Northern Burma it is very common, and known as *Hintha*. The Hindustani name *Chakwa* (with the feminine form *Chakwi*) is not the only one, *Surkhab* being also used; *Nir-batha* or *-koli* is the name in South India, *Mungh* in Sind, *Bugri* in Bengal, the Telugu name is *Bapana Chilluwa*, and the Marathi *Sarza* or *Chakrawak*.

1/3

TADORNA VULPANSER

Common Sheldrake.

Tadorna cornuta. *Shah-Chakwa*, Hindustani.

The real original or typical sheldrake, a well-known sea-coast bird at home, and the only surface-feeding duck which is a sea-bird anywhere, is a rather rare winter bird only in India, being only at all common in Sind, where it is called *Niruji*, and not going far south anywhere, though it ranges east to Upper Burma. The Hindustani names of *Safaid Surkhab* and *Chandi Hans*, however, which are in use as well as that given above, show that the natives know the bird well, and it is one that once seen is never forgotten—its predominant white colour, indicated by its native names, and set off by a black head and wing-tips, chestnut breast-band joining on the shoulders, and scarlet bill, are quite unique and unmistakable. Even the yearlings, in which there is no chestnut tint, and whose beaks are merely flesh-colour like the feet, are quite unlike any other duck. In size this bird is a little less than the ruddy sheldrake or Brahminy duck, being about as large as the mallard or spotted-bill, though much higher on the legs.

It walks and runs well and gracefully like its ruddy relative, and also swims high in the stern; the male floats particularly high in the water and looks decidedly bigger than the female, but there is practically no difference in plumage, although the drake's is richer in its hues. He has, however, a knob at the base of the bill in the breeding-season, and some trace of this is always visible in fully adult birds.

The note differs greatly in the two sexes in this sheldrake, being in the male a low whistle, while the female's is loud and harsh, something between a quack and a bark. Though perhaps more often seen ashore than afloat, this duck is more of a water bird than the Brahminy, and can at a pinch dive well and go some distance under water. It is wary and hard to shoot, and as food it is one of the very worst of ducks, and indeed is not usually regarded as eatable. It feeds chiefly on small animal life, especially minute shell-fish, but also eats grass. All across the Old World it is a well-known bird in the north by the sea and lakes, but the bird known as sheldrake in the United States is our red-breasted merganser; the prefix "shel" means pied, and no doubt was originally common to both species, as showing much white.

White-winged Wood-Duck.

* *Asarcornis scutulata.* *Deo-hans,* Assamese.

The white-winged wood-duck is easily distinguished from all our other water-fowl by the contrast of its white head and the white inner half of its wings with its entirely dark body; its great size, which exceeds that of all other Indian ducks, is likely to cause it to be mistaken for a goose when seen on the wing at a distance, but close at hand, whether seen on land or water, it is a most unmistakable duck, with nothing of the goose about it.

The plumage of greenish-black and dark olive and red-brown, the black speckling on the white head, and the unique blue-grey bar bounding the white of the wing are common to both sexes, as are the yellow of the bill and feet, the former more or less speckled with black; but it is only in the male that the bill becomes red and swollen at the root when the bird is in breeding condition, and he is very noticeably larger than the female, which, big bird though she is, does not average more than five or six pounds; a drake weighs about eight.

This splendid duck is a resident in our Empire, but very local and even yet not well known to most people, although the investigations of Mr. E. C. S. Baker in recent years have taught us a good deal about it. Its main home appears to be Assam, but it ranges east through Cachar and Burma to the Malay Peninsula, and its great haunts are the jungly, marsh- and pond-studded tracts in the country at the base of the hills; in any case they are to be looked for in forest pools and streams, provided the running water is sluggish. It will thus be seen that their haunts are different from those of ducks in general,

---

* *Casarca leucoptera* on plate.

CASARCA LEUCOPTERA

and in a suitable locality a couple of brace may be got in a day, not, of course, without considerable exertion. The birds spend much of their time on trees, and generally occur in pairs or even alone; flocks do not generally number more than half a dozen when met with.

Although so easily tamed that except in the breeding season they can be allowed liberty and even the use of their wings, they are very wary and hard to get near in the wild state. The flight and call are described as goose-like, the note being a loud squawking or trumpeting; nothing is said about there being any sexual difference in the voice, nor does the male's courting behaviour appear to have been recorded.

Those Mr. Baker kept do not, indeed, seem to have shown much inclination to breed beyond pairing regularly, and he found them remarkably good-tempered. This is probably a sign that they were never in real high condition, for birds that go in pairs ought normally to want to "clear the decks" when they think of nesting. Hume similarly found Brahminy ducks very gentle and amiable, whereas, as I have said in my account of that species, they are really quite the reverse if determined on domesticity. A single female wood-duck in the London Zoo recently was sluggish in her habits, and quiet when with Muscovy ducks, but when among the smaller water-fowl I have seen her make a spiteful grab at one now and again. I noticed that the gait on land and style of floating in the water, in this bird, were not in the least like those of the nukta or sheldrakes, with both of which this species has been associated, but like that of an ordinary duck such as the mallard or spotted-bill. The squawking voice, goose-like flight, style of wing-marking, and general habits, however, seem to point out that this bird is really a peculiar type of sheldrake, and the swelling of the drake's beak in spring is similar to what happens in the common sheldrake, as well as in the nukta.

Although Mr. Baker's specimens dived freely to catch live fish put in their tank, the wild birds are found not to dive when wounded, but to go ashore and hide in the jungle. They like various small animals, such as snails, insects and frogs, as well as fish, and prefer these to grain in captivity, though they would eat and thrive on the latter; they would not touch dead animal food, which is curious, as mergansers make no difficulty about this, though true fish-eaters. I presume they are fairly good eating, as an Assam planter who shot them regularly used to eat them equally so. The birds nest in holes and hollows of trees, the breeding season being about May, and they moult in September, retiring to the most remote swamps for safety. Outside our limits this bird is found in Java, and is said to be domesticated there.

Small Whistler or Whistling Teal.

*Dendrocycna javanica.** *Silli*, Hindustani.

The loud whistling call of several syllables uttered by this duck will at once strike the newcomer to India as something new in duck utterances; it has evidently given the bird the Hindustani name above-noted, as also the variants of *Silhahi* and *Chihee*, while the Burmese rendering *Si-sa-li* is even closer.

The flight of the bird is as distinctive as the note; the legs, unusually long for a duck, and the neck tend to droop, at any rate when flying low, and the large blunt wings, which are all black underneath, contrasting with the brown body, are moved quickly, although the flight is not fast. The birds may settle in a tree, which naturally seems an even more remarkable performance than their vocal one; and, indeed, the *dendrocycnas*, which are essentially tropical ducks, are often called tree-ducks as a group. The whistling, cackling call, however, which is common to both sexes, is a far more distinctive peculiarity than the perching habit, common as this is to most ducks resident in the tropics. No doubt crocodiles and alligators have done their share in establishing this custom!

The plumage is as alike in the two sexes in this duck as the call, and this is again a group peculiarity; in the present bird there is no striking marking, but the combination of

* *arcuata* on plate.

DENDROCYGNA ARCUATA

brown body and wings nearly all black will distinguish this bird from all our species but the large whistler, of which more anon. On the water it swims rather low, and the neck seems long in proportion to the narrow body and very short tail, while the wings fold so closely that the tips are not seen. This is a small duck, only weighing a pound or a little over, but it is absurd to call it a teal on that account; the teal are pigmy relatives of the typical ducks, while these whistlers are a very distinct group, and in many ways are more like small geese than ducks. The present bird is the most abundant of the resident Indian ducks, and is found nearly all over the Empire where wood and water are combined, even down to the Andamans and Nicobars. But it is essentially a warm-climate bird, and does not often ascend the hills, nor is it to be found in dry treeless districts. In the Punjaub, where the migratory ducks are so common, it is rarely seen.

The sort of water it likes is that overgrown with weeds, and here it is quite at home, be the water a village pond or an extensive *jheel*. At night it roosts on a neighbouring tree, feeding among the weeds during the day, but seldom going ashore to do so. On land it walks well and gracefully, though slowly; but it is essentially a water bird, and dives for food freely, though its action in so doing is just like that of a coot, as it springs high in starting, lifting its whole body out of the water. Naturally, it is difficult to bring to book if wounded, but Europeans generally refuse to regard it as game, owing to its general tameness and slow flight. This is a mistake, for it occurs in flocks of thousands where the locality suits it, and where it is common it must greatly interfere with the game migratory ducks. It is a most quarrelsome bird with others, attacking in combination; I have seen even four set on to one spotted-bill in captivity. Even its own big cousin next to be described comes in for its bullying, and gives way to it. The flesh of whistlers is poor in most people's opinion, but will do for soup, and is liked by natives. The food is water-plants and snails, rice, &c.

The birds breed usually in holes of trees or in the old nest of some kite or crow, but they also make nests for themselves either in trees, on cane-brakes, or among rushes—about anywhere where any duck ever does nest, in fact, except underground, though nests made by the birds themselves on the boughs, a rare habit among ducks, are the most usual. The eggs are rather rounded, very smooth, and creamy-white when fresh; while the female is sitting the drake keeps guard close by, and the young are carried down to the water in the old birds' feet. The bird, which extends outside our limits to Java, has many native names: *Saral, Sharul, Harrali-hans,* in Bengali; *Hansrali* in Uriya; *Horali* in Assam; and *Tatta Saaru* in Ceylon; *Tingi* in Manipur; the Telugu name is *Yerra Chilluwa.*

Large Whistler.

Dendrocycna major. *Burra Silli,* Hindustani.

The large whistler presents the peculiarities of its small common cousin in an exaggerated form; it is longer-necked and more leggy, and has bigger feet and head; its wings are blacker, and its body-colour a much richer brown—chestnut instead of dun, in fact; its feet are often much lighter, French-grey instead of dark slate; and it is far bigger, weighing up to two pounds, though it takes a good male to reach this. Nearly all these distinctions, however, though the most striking, are comparative; more positive ones are the presence of a transverse curved patch of cream colour above the large whistler's tail, most noticeable when it takes wing; this is replaced by dark inconspicuous maroon in the small whistler, which, on the other hand, has a conspicuous yellow ring round the eye, owing to the edge of the eyelids being thus coloured; in the large bird they are grey, like the bill and feet. On the water the much more strongly developed streaking of cream colour on the flanks, as well as the redder head and breast and darker back, make the big whistler noticeable.

It swims as well and dives as freely as the small kind, but also comes ashore a great deal more, and does not divide its time so rigidly between the water and the trees. On the wing it is far swifter, and being more wary, is a really sporting bird;

1/3
DENDROCYGNA MAJOR

some people also think it better on the table. It is resident in our Empire, but cannot be called a common, widely distributed or abundant bird; it only goes in small flocks, never in the dense masses such as are seen in the case of the small whistler, and is only really numerous in Bengal, though it ranges east into Burmah, west to the Deccan, and south to Madras. Outside our Empire it is not found in Asia.

It feeds on much the same food as the small whistler, with an especial fondness for rice, wild or cultivated, and selects the same situations for nesting as a rule, *i.e.*, old nests, holes in trees, or suitable boughs on which the birds make a nest of their own; they have not, however, as yet been found nesting on the ground, but this is probably because they are so scarce and local in comparison that there are not the same opportunities for observation, or for variation in the birds' habits for that matter, that there are in the case of the small common species.

Although they are afraid of this bird, and less aggressive with other ducks, they will fight readily enough with each other in captivity, springing right out of the water and striking with their feet. They also pair freely, unlike most of this group, which display little sex-proclivities in captivity, except for tickling each other's heads like doves or love-birds. The eggs are white, and rather larger than those of the common whistler, though extremes meet.

Outside Asia the large whistler, sometimes called the fulvous duck, is found in tropical Africa and the warm regions of America. This is a most extraordinary range for a bird that does not undertake long migrations, and calls for considerable elucidation; there can be no doubt that the bird is greatly disadvantaged in India by the competition of its abundant and aggressive relative, but then, on the other hand, it is far more hardy, bearing the English winter outdoors when the small kind looks thoroughly miserable and soon dies off, and even breeding. So one would think that it might have colonized cooler climates and struck out a line of its own; but possibly it once had a more northerly range, and has become reduced to its present location by some cause which we do not at present understand—at any rate, its persistence in indistinguishable form all round the tropics is a unique phenomenon in bird-life.

Bar-headed Goose.

Anser indicus. *Kareyi Hans,* Hindustani.

Although not breeding in India, but a winter visitor only, this is *the* wild goose of the country, visiting it in enormously greater numbers than any other species, and being far more widely distributed. The white head marked with two black cross-bars is unique among geese, but this colouring is not found in the young of the year, which have the crown brown continuous with the back of the neck. The real and most striking peculiarity is the pure light grey colour, more like that of the ordinary gulls than the usual brownish grey of geese in general: the legs are orange, and the bill the same or lighter, black-tipped. These geese seldom weigh quite six pounds.

The bar-headed goose is commonest in Upper India, but is not plentiful in the Central Provinces, and decidedly rare further south. To the east it is common in Upper Burmah, and ranges into Manipur, where it is called *kang-nai.* Ceylon it does not visit at all; none of the true geese occur there, in fact, all being essentially birds of the north. From Gujarat it is also absent, but at the opposite end of India, in Western Bengal, extremely numerous.

Like geese generally, it is eminently a gregarious bird, and the flocks are sometimes very large; they may contain as many as five hundred birds. Hume says that he has seen as many as ten thousand, in flocks of varying sizes from one hundred up, on a ten-mile reach on the Jumna. Of course large birds like these, in such numbers, do an enormous amount of damage to crops, all sorts of herbage, whether of pulse or grain, coming into their bill of fare, though late rice is perhaps the favourite. As they commonly feed at night, though when undisturbed they will graze up to 9 a.m., and long before dark, a great deal of harm can be done without much chance of its being averted. After a course of this sort of feeding, they are in fine condition for the table at the appropriate time of Christmas; but when they first come in, in October, they are thin and in poor case. As a general rule they all leave for the north again in March or early April. This goose prefers rivers to standing water for

ANSER ¼ INDICUS

its daytime rest, but, like the geese generally, does not go into the water much, but remains on the banks, with sentries set to give the alarm if required. Sometimes, however, the flocks ride at anchor, as it were, in the middle of a river or tank.

Drifting down on them in a boat, where they are found on a river, has been found a satisfactory manner of approach; and they are often shot at flighting-time in the evening. They fly in the V-figure which is usually assumed by travelling geese, and, for such birds, are unusually active, as well as strong, on the wing. Damant records a curious example of this: "In Manipur," he says, "I have often watched them returning from their feeding grounds to the lake where they intend to pass the day; their cry is heard before they themselves can be seen; they then appear flying in the form of a wedge, each bird keeping his place with perfect regularity; when they reach the lake they circle round once or twice, and finally, before settling, each bird tumbles over in the air two or three times precisely like a tumbler pigeon." I have seen domestic geese turn somersaults in the water when playing, but "looping the loop" is somehow a performance one hardly expects of a goose. In spring, also, in its breeding haunts, the bird chases its mate on the wing. As geese go, however, this species is graceful and active on land also; it sits high in the stern on the water, like geese generally. The note is harder and sharper than that of the grey goose, according to Hume, who says the two species can be distinguished by this alone when passing overhead at night. They do not associate, though often seen near together.

Although the bar-headed goose is found on the Kashmir lakes and elsewhere in the hills up to 7,000 ft., it does not breed in India, but in Ladakh, Central Asia and Tibet. The four or five eggs are white, and are to be found on islands in the Tso-mourari lake even before the winter ice breaks up. The goslings are yellow, shaded with olive above, and have black bills and feet, as I have seen in specimens bred in Kew Gardens. The Ladakhi name for the bird is *Neg-pa;* the Nepalese call it *Paria*, and the Tamils *Nir-bathu; Birwa* is also a Hindustani name.

Grey or Grey-lag Goose.

*Anser ferus.** *Sona hans*, Hindustani.

Everyone who has seen a grey domestic goose at home knows what this bird is like; only the wild race is smaller, its form is slighter and more elegant, and the beak and feet, generally orange in tame geese, are pink or flesh-coloured. In spring the bill becomes very rich in tint in Indian specimens, a bright rose or light carnation red.

Old birds are heavily marked with black on the belly, and these should be avoided when selecting geese from one's bag for one's own consumption, according to Hume's sage advice, as apt to be tough and hard. Such birds may weigh as much as eight and a half pounds. .On the wing this goose can be discriminated from all the other dark-grey or brown species by the pale French-grey tint of the inner half of the wing, which shows up very conspicuously in flight, appearing nearly white. The gaggling note is like that of the tame goose at home, but not so shrill and high as that of Indian tame geese, which are of the Chinese species (*Cygnopsis cygnoides*) so well known as ornamental birds in our parks. This black-billed brown goose is found wild in Eastern Asia, and may hereafter be found to occur in the east of our Empire in that condition.

The grey goose is the only common goose in India besides the bar-headed, and, like that bird, is only a winter visitor; all along the northern Indian and Burmese provinces it is common, but its numbers bear no comparison to those of the bar-head except in Sind; in Gujarat, however, it is the only kind found. Like the bar-head, it visits Kashmir and parts of the Himalayas at a moderate altitude. Its southern limit for the most part is that of the Gangetic plain.

It is, if anything, more gregarious than the bar-headed goose, flocks of upwards of a thousand being seen on the west, where it is most abundant; these flocks in flight observe the usual V formation and travel with a rapid but stately flight. They get under way slowly, and Mr. E. C. S. Baker advises that when

---

* *cinereus* on plate.

ANSER CINEREUS

stalking them one should put in one's first barrel at them on the ground, and give them the second as they rise. Although wild geese are often much less wary in India than they proverbially are in Europe, they will be found to need careful stalking where natives have guns, and in such places it is of no use getting one's self up as a native in a blanket disguise, a bullock used as a stalking-horse being much better.

They may also be shot when by the side of rivers by gliding down on them in a boat, as mentioned in the case of bar-headed geese, but there must be some arrangement to conceal the shooter's head. They keep more on the shore than in the water, and walk well, if not so gracefully as the bar-heads; they are also fast swimmers, and dive freely in play or when wounded, but cannot keep under long. Having the same vegetarian habits as geese in general, and being often so numerous, they are only second as crop ravagers to bar-headed geese, and like them, do much of their mischief at night. The younger birds, when well fed, are good eating; actual yearlings may be distinguished by having the feathers of the usual rounded shape, the square-tipped feathers being a peculiarity of geese after they have got their adult plumage, and particularly noticeable in the darker species owing to the light tippings showing up in transverse bars on the back and flanks.

Even with big birds like this goose, however, eagles, and in tidal waters crocodiles, prove a great nuisance by making off with wounded birds, astonishing as it may seem that a comparatively slight-built eagle like the common ring-tailed river eagle of India (*Haliaetus leucoryphus*) should be able to lift and carry such a weight, which must much exceed its own. Grey geese come in and depart at about the same time as the commoner species; their breeding-grounds are in Northern and Central Asia and in Europe, including a few localities in Britain. They also visit Europe in winter, but at home are the least numerous of the regular visitants among the geese. It is just possible they may be found breeding in Kashmir; the nest is a mass of reeds, &c., piled upon the ground near water, and the eggs white, and about half a dozen in number. The goslings have black legs at first. In Hindustani this species is sometimes called *Raj-hans* as well as the bar-headed, and similarly *Kangnai* is used in Manipur. The Nepalese name *Mogula* is, however, quite distinct.

Pink-footed Goose.

Anser brachyrhynchus.

Although Hume noticed in a pair of these birds he shot from among a flock of greys in 1864 on a sandbank in the Jumna, that as he looked down upon them from a cliff above "they were conspicuous by their smaller size, clove-brown colour (that is what they looked at a distance), and very pink feet," some allowance must be made for variation. The feet of the grey-lag are sometimes about the same shade as those of the pink-footed, and the tone of the plumage also varies, as well as the size, the pink-footed sometimes weighing as much as the smaller grey geese.

The real distinguishing point is the bill, which is black at the base and tip, only pink on the intermediate space, in young birds only a band nearer the end; the pink is sometimes very rich—carmine, in fact—and sometimes, both on the bill and feet, verges on or is replaced by orange. The slaty-grey inner half of the wing, however, which resembles the same part in the grey goose, though darker like the rest of the plumage, will distinguish a pink-footed goose whose feet are not a true pink from the orange-legged bean-goose; besides which the bean-goose is a big-billed bird, the beak being two and a half to three inches, while the pink-footed, as indicated by its scientific name, which means "short-billed," has a particularly small beak, only a couple of inches long, and narrow in proportion.

The pink-footed goose is one of our winter rarities, but has been reported on several occasions, though actual specimens have very rarely been preserved. It has been reported from the Punjaub, Oudh, Assam, and the Shan States, and Mr. E. C. S. Baker has one procured in Cachar. This specimen was got by a fluke by one of his native collectors out of a very wide-awake flock of about a dozen, and a flock of twenty has been reported from the Punjaub.

W. Foster

ANSER BRACHYRHYNCHUS.

It is possible, however, that, as Mr. Baker says, this species may have been mixed up with Sushkin's goose, *Anser neglectus*, a little-known species recently described, ranging as near as Persia, and believed to have been obtained in India.

This also has pink feet and the centre of the bill pink, but it resembles the bean-goose in plumage, having no slate-colour on the wing, and its bill is about two and a half inches long, though not so well developed as in the bean-goose. I must say I am very suspicious of all these supposed new species of geese, and I am inclined to suspect that as the pink-footed goose may have orange where it ought to be pink, the bean-goose may return the compliment. Also, what is there to prevent these nearly allied geese from hybridizing, which geese do most readily in captivity? Anyone, therefore, shooting a pink-footed goose of any sort should be careful particularly to record the colour of wing and size of bill, these being the great points in these half-black-beaked species of geese.

The true pink-footed goose is a western bird, its breeding-places being in arctic Europe; it is the commonest of the "grey" inland-feeding geese known at home—in fact the most numerous of all our wild geese, except the "black" brent of the sea coasts.

Bean-goose.

*Anser fabalis.**

The bean-goose, like the grey-lag, is a big bird with a large strong bill, indeed bulged and coarse in some specimens; by the colour of this, which is black at the tip and for a varying amount of the base, the remaining portion being orange, it can be distinguished from any large-billed goose found with us. The legs are also orange, and the general colour of the plumage dark greyish-brown rather than grey.

Although there are several reports of its occurring in India, there seems to be no actual Indian-killed specimen on record; yet Blyth, who was perhaps the best naturalist that ever lived,

* *segetum* on plate.

definitely said that Gould had one which was killed in the Deccan.

As the bird has a wide distribution in the north of the Eastern Hemisphere, and visits the shores of the Mediterranean in winter, it is at any rate highly probable that these records are correct. Several forms are distinguished, chiefly differing in the distribution of the colours on the bill and in the size. The bird figured in Hume's plate is of the race distinguished as *aroensis*, in which the orange occupies the bill nearly to the exclusion of the black, whereas in the typical form this colour is confined to a band in front of the middle of the bill, the bill being thus mostly black. This is generally the case also with the race *middendorfi*, which visits China in winter, and is certainly a probable visitor; but Blyth said that Gould's bird was a common bean-goose, while the bird named after von Middendorf is particularly large, especially with regard to the bill, and is prone to exhibit a yellowish shade on the head, admittedly variable, however, and accidental, like the rusty stain so often found on the head of the swan. It is also admitted that there may be much more than a mere band of orange on the bill of this Eastern bean-goose, and so it seems most advisable at present to call them all simply bean-geese; anyone who shoots a specimen can indulge in details of marking and measurement to his heart's content.

White-fronted Goose.

Anser albifrons.

Every goose that has some white on the forehead is not necessarily a white-fronted goose, nor does the absence of white in that part disqualify a bird for the title; for the grey-lag often shows a little white at the base of the bill, and young white-fronted geese have none, but are rather darker there than elsewhere.

The real distinctive points are the combination of orange legs with a yellow or flesh-coloured or pink bill, changing after death to orange, but in any case without any black on it. The body-colour of this bird is much browner than in the grey goose,

1/4
ANSER SEGETUM

and the size much smaller as a rule, large specimens weighing little over five pounds. The beak is also smaller in proportion, measuring only two inches, while the grey-lag's is two and a half to three.

The white front, when fully developed in adult birds, is in the form of a broad band across the forehead and bordering the base of the beak; in adult birds also the belly has transverse black markings, often so pronounced that this part is practically all black.

The white-fronted goose is with us a rare winter visitor, but may turn up anywhere in our northern territories, from Sind to Burma. It is met with by itself, not along with other geese, although only found singly or in twos and threes. Three, for instance, was the number observed and shot by Hume on his first record of the species in November, 1874; these birds were shot on the Jhelum, and one of them, which was only wounded, led him a literal wild-goose chase before being secured, twice rising and flying off strongly for a long distance before he finally got it, to the great disgust of his men, who, as he says, "were tired of plodding through the loose sand; all objected to going further after this goose. In the first place they declared he had flown away altogether out of sight; in the second place they said I might have killed a dozen geese during the time I had wasted over this one wounded bird, which was, moreover, a very small one. There was almost a mutiny, but I had marked the bird precisely and insisted on going up to the spot," when the deplorable disregard for science exhibited by these benighted natives was punished by the goose's opportune appearance just as it was despaired of even by their master.

These specimens proved to have "fed entirely on some species of wild rice, and on tender green shoots of some grass or grain"—the ordinary food of geese, in fact. Indeed, there is nothing special about the habits of this species to record, though its cry is rather different from that of the grey goose, being often compared to laughter; in fact, laughing goose is a well-known name for it. When in India it has usually been found to frequent rivers. It ranges all across the Northern Hemisphere in high latitudes, though the American birds, which are larger than the Old World ones, are nowadays commonly referred to a so-called species, *Anser gambeli.* Should this large, and especially large-billed, form turn up here, the entirely light bill will distinguish it from all our large species except the grey-lag, and the orange feet from that. It goes as far south as the Mediterranean in winter, and at that time is one of the familiar wild geese that worry farmers and bother shooters at home.

Dwarf Goose.

*Anser erythropus.**

This small edition of the white-fronted goose is hardly bigger than the Brahminy duck, and being of a decidedly dark brown colour, is recognizable at a considerable distance. Close at hand, it will be noticed that the eyelids are of a yellow colour, forming a noticeable ring round the eye; this forms a positive distinction from the large white-fronted goose, and besides, the bill is small even in proportion to the smaller size of the bird, as usual in small geese, while the wings are longer, reaching even beyond the end of the tail. It is not, therefore, surprising to hear that it is more active in flight than the big geese.

These points will distinguish even the young bird before the white on its forehead has developed, though all the Indian specimens I have seen have had it; and this white patch when present is another good distinction, being much longer and extending up on to the crown, instead of developing mostly in a transverse direction. The legs are orange as in the large species, and the bill is said to be orange in the adults and reddish grey in the young, but in those live specimens I have seen, mostly fine adults, the bill has been a very bright rose-pink or cerise red, though one or two, no doubt younger, certainly so in one case, had it flesh-colour. I expect the colour changes after death, or is individually or locally variable, as in the grey goose, which in Europe has often an orange bill.

The dwarf goose is only a winter visitor to India, and if anything rarer than its larger relative; very few have turned up

* *minutus* on plate.

W. Foster

ANSER ALBIFRONS

since the first record in October, 1859, when Irby killed a couple and saw another near Sitapur in Oudh. But it has been got here and there in the north as far east as Lakhimpur. In 1898, four came into the Calcutta Bazaar, and I got them on behalf of the Calcutta Zoo; three were brought all together on New Year's Day, but they had been sent down from up-country, and had their wings cut. It is worth noting that these birds did not moult at the proper time that year; two died, and I determined to pull the quills of the others to start the moult and save their lives. This I did, not without difficulty, but the result was they moulted all right, and lived and moulted normally for some time after, not seeming to feel the heat at all, though this species is a high northern bird, breeding close to the unmelting ice. It is common in winter in China, and also visits Japan, but not any part of the New World. It is an Eastern species, and as rare at home as in India.

Red-breasted Goose.

Branta ruficollis. *Shak-voy*, Siberia.

A very small black-and-white goose, with a breast nearly as red as a robin's, is such a remarkable bird that it can hardly be overlooked anywhere, and so the fact that it has been recognized in India is not surprising, though the paucity of even visual records, and the absence of any actually obtained specimen, bear eloquent witness to its rarity.

Like the dwarf goose, it is hardly larger than the Brahminy duck, and has very long wings and a particularly small bill; its colours also, curiously enough, are practically those of the Brahminy, though very differently distributed. The general hue is black, with white stern and broad white band along the flanks; as this comes just above the water-line, the bird would not look nearly so dark on the water as ashore or on the wing. The rich reddish-brown of the fore-neck and breast is also bordered with white, as is a patch of the same red on the cheek; before the eye there is a white stripe.

One would expect the female of so richly coloured a bird to be at least a little duller than her mate; but this is not the case, the rule among the true geese of the similarity of the male and female being strictly observed, and the sooty-brown specimens with dull brownish tints where the red ought to be being the young of both sexes. Even these, however, are quite unmistakable. The bill and legs are black, the former being remarkably delicate and small, only about an inch long.

The best record of the occurrence of this lovely bird in India is that furnished by Mr. E. C. S. Baker, in his book on the Indian ducks and their allies; he says there that he "was fortunate enough to see five specimens on a chur in the Brahmapootra, just below Gowhatty; they arose a long way off as the steamer drove up stream towards them, but turned and flew past us within sixty to a hundred yards, and there could have been no possible chance of mistaking them." His friend, Mr. Mundy, had previously communicated to him a good description of some he had seen on the same river in Dibrugarh. As to the record of 1836 in the *Oriental Sporting Magazine,* I have looked this up, and quite agree with Blanford that the author of this did not know what he was talking about; so that these modern ones, in my opinion, remain unique. The red-breasted goose occurs in Europe, including England occasionally, but always as a rarity; it is, however, not really a very rare bird, being common enough in Western Siberia, where it breeds, and coming as near to us in winter as Persia and Turkestan. It does well in captivity, and while this book was being written I had the pleasure of inspecting a lovely pair which were deposited at the Zoo *en route* from Germany to the Duke of Bedford's estate at Woburn, where, I heard, there was already another. These birds showed the tame disposition with which this species is credited; and I must say that if I got hold of a netted or wing-tipped bird in India, I should not dream of killing it, but keep it to send to Europe, since a photograph, if only of the head, would be amply sufficient for the record.

1/4

ANSER MINUTUS

Mute Swan.

Cygnus olor. *Penr*, Punjabi.

The only swan which visits India in any numbers, and that only in hard winters, is the well-known bird that is kept as an ornament all over the civilized world. No doubt a few come in every winter, and they have been killed in the hot weather on two occasions; but that the bird has always been a rarity is proved by the fact that Calcutta dealers have for many years imported them from Europe by the dozen, and by the fact that there is no true native name—*Penr* really meaning a pelican.

This swan may be distinguished from all others by the black knob at the base of the bill, but as this is little developed in the young birds, the best point to go by is the colour of the bare patch, which extends from the bill to the eye; this in this species is black as well, whatever the age. In young birds the plumage shows more or less drab, and their bills are not of the full orange-red colour of the old birds, but some shade of grey or pink.

Although so well known as a tame bird, and well established as an "escape" breeding at large in some parts of Britain, and, doubtless, elsewhere, this swan has, for a water-bird, not a very wide range; nor does it go very far north, its true home being Central and South-eastern Europe and Western and Central Asia. In winter it visits North Africa, but does not go very far west; and India appears to be its eastern limit on its southerly migrations. And with us it only comes to the North-west, the Peshawar and Hazara districts being the most likely ones in which to find it. The birds have generally been seen singly or in small flocks, and have shown a tameness which has been rewarded by unrelenting slaughter in too many cases—as if one such bird were not enough for a record, the species being so unmistakable.

At the same time, although swans are but rarely eaten in Europe nowadays, it may be remembered that they are edible—at any rate the grey yearling birds—which are still fattened for eating at Norwich, if nowhere else in England. In view of the occasional occurrence of these swans in summer, and of the fact that they have laid eggs when kept in captivity in such an unnatural climate as that of Calcutta, it is just possible that they may yet be found breeding somewhere in India, especially in exceptionally cool seasons. Most people know what a swan's nest is like—a huge pile of any vegetable matter the birds can get hold of, placed close to the water's edge, and, if possible, on an islet. But as the birds, to put it mildly, do not encourage examination of the nest when without fear of man, it may be as well to mention that the eggs are about four inches long, pale sage-green in colour, and number about half a dozen. The cygnets are grey normally, but now and then white ones occur; and these are white even in their first feathering, and have pale clay-coloured or flesh-coloured feet all their lives instead of the usual black or grey. Such birds used to be distinguished as a species, the so-called Polish swan (*Cygnus immutabilis*).

The food of these swans consists of water-weeds and grass, with some animal matter, especially fish-spawn; in domestication they eat grain freely, but do not come ashore to seek it in the wild state apparently. In fact they do not come ashore much except to rest, generally grazing from the water, where grass on the banks is accessible; nor, though they stand on their heads to reach the bottom, do they ordinarily dive; though I once saw a small cygnet do so for about a couple of yards when attacked by a vicious black swan. This Australian bird, by the way, is more freely imported into India than the mute swan, and both species have been known to escape; so that records, especially if of old birds, and away from the North-west, are not free from suspicion. The birds rise heavily and slowly, but fly fast, though with slow strokes, and, in spite of their awkward gait, a wounded bird has been known to run fast in hundred-yard spurts before hunted down.

This species is well called the mute swan, for though not actually voiceless, it is far more silent than other species, and its note, a grunt or a sort of suppressed bark, is not loud. It is one of the largest of flying birds, attaining a weight of thirty pounds; though the birds occurring here are not likely to weigh more than half that.

CYGNUS FERUS. CYGNUS OLOR.

Whooper.

*Cygnus musicus.**

The whooper, often distinguished at home as the " wild swan," is a far rarer visitant to India than the mute swan, having been recorded in India less than half-a-dozen times. The earliest record was, curiously enough, in Nepal, and many years before the first record of the mute swan, namely, in 1829.

All the other specimens have been got in the North-west, more than one having sometimes been seen.

This swan is not noticeably smaller than the other, and is also drab in the first feathering, though white in the down as well as when adult; but it is easily distinguishable at all ages, because the bare patch of skin on the face is always pale, not black—greenish-white in the young, and bright yellow in the adult. The end half of the bill, or less, is black, the old one's beak being black up to the basal end of the nostrils above, though the black does not reach beyond the further end of them below; the rest of the bill is yellow, continuous with the yellow face.

The real difficulty is to distinguish this bird from Bewick's swan, whose distinctions, however, are given below. The bill has no knob, and is longer than in the mute swan, while the nostrils are situated farther forward, being in the middle of the bill, while in the other bird they are nearer the base than the tip.

Other distinctions are the short blunt tail, the mute swan's being pointed; the straight goose-like carriage of the neck, and especially the voice, which in this species is a beautiful trumpet-call. This is evidently the swan celebrated in ancient story as singing before its death; in fact, one bird shot in India, on the River Beas, being only winged, "continued to utter its long, loud, musical trumpet-call," while the three birds which had accompanied it were still in sight, as recorded by General Osborn, who shot it, in a letter to Mr. Stuart Baker.

The whooper is a true northern bird, being found in the

---

* *ferus* on plate.

breeding season chiefly in the Arctic Regions, both in Europe and Asia; but it breeds as near and as far south, apparently, as Seistan, and also nests in Greenland, though not on the American continent. In Iceland it is well known as a nesting species. In winter it regularly comes as far south as Southern Europe in the West, and Corea in the East.

Its general habits are similar to those of the mute swan, but it comes ashore to graze more, and is not so awkward a walker.

Bewick's Swan.

Cygnus bewicki.

Bewick's swan is so very like the whooper that it requires a fairly near view to differentiate them, for though Bewick's is a considerably smaller bird, this cannot be appreciated unless there are facilities for comparison, and dimensions vary in both species; while of course detailed examination is necessary correctly to observe the distribution of black and yellow on the bill, which furnishes the most reliable distinction. Bewick's swan, like the whooper, has the face and base of the bill yellow, but the yellow is confined to this part of the bill, all the rest being black from the nostrils to the tip, as is also in some cases the ridge of the bill between the nostrils and the forehead. The yellow generally stops short at or before the basal end of the nostrils, all across the beak, and never extends below the further end of them and even beyond that, as it does in the whooper. The young birds in this species are grey, and have flesh-colour on the bill where the old ones are yellow; the weight is up to twelve pounds, little smaller than some whoopers. The first undoubted Indian example of Bewick's swan was recorded by Mr. E. C. S. Baker, in the *Bombay Natural History Society's Journal* in 1908, vol. xviii. It was a fine adult bird, and had been killed at Jacobabad, by Mr. McCulloch. In the winter of 1910-1911, two more specimens turned up, one near Mardan, and another at Campbellpur, on December 30 and January 2 respectively; both were adults apparently, and the exceptional cold then prevailing no doubt, as Mr. Baker suggested in record-

1/7
CYGNUS BEWICKII.

ing these specimens, had caused their appearance in our limits; at the same time he recorded the occurrence of a couple of young whoopers, shot out of a flock of seven on the Kabul River. It is very possible that a swan recently recorded as seen near Bhamo, though not bagged, may have been of this species, as it is said to have had a small black bill. In any case, there is now no doubt about the occurrence of Bewick's swan as an occasional visitor to India, while probably Burma also is within range of its winter wanderings. In fact, as this bird has a more eastern range than the whooper, at any rate in the breeding-season, it might be reasonably expected to come in at least as often as that species; its normal winter quarters in Asia, however, are China and Japan, and in China it seems to be the commonest swan at that season. In the west it ranges in winter as far south as the Mediterranean.

It has a quite different and less musical note than the whooper, resembling the syllable "*kuk*" many times repeated, and sits high on the water. It comes ashore a good deal, and is a better walker than most swans; it can, moreover, run well.

Common Snipe.

*Gallinago cælestis.** *Chaha*, Hindustani.

The "fantail" snipe, as this species is often called to distinguish it from the next, is the same bird as the common snipe of Europe; I mention this particularly, because I have heard of sportsmen proposing—and I believe the idea was carried out—to send some Indian-shot snipe home in cold storage to see if they really were the same as the British birds.

But to many people who have not done much shooting before they come out, the difficulty will be to distinguish a snipe from the many sandpipers or snippets; such small waders being abundant in India, and often sold—at any rate they were in

* *scolopacinus* on plate.

the Calcutta Bazaar in my time—for the table as snipe, which argues that a good many people do not know what a snipe ought to be.

The characteristic of the true snipe, fantail or pin-tail, then, is the rich dark, well-mottled plumage, showing brown, black and buff, instead of the more uniform and coldly tinted drabs and whity-browns of the sandpiper tribe. In particular is to be noticed the orange-buff tint at the end of the tail-feathers. The great length of the beak, which is about half as long as the rest of the bird, is also a noticeable point, but as some few sandpipers have very long bills, it may be noticed that in such there is always a small web between the middle and outside toes, which is completely wanting in snipe of all kinds.

There is no difference in colour between male and female snipe, but, on the whole, the hens are bigger than the cocks, the hen's bill sometimes reaching three inches, and her weight five and a half ounces, while a big cock's beak will only be about two inches and three-quarters, and his weight barely over the five ounces; the average weight of both sexes is given by Hume as 4·2 ounces.

The vast majority of fantail snipe are winter visitants to India; they first come in in any number about the end of August, and September is the usual month for the arrival of these birds, while in Southern India and Burma they are later than this. By the end of March most of them have usually left again, but sometimes many stay on till the middle of April, and even up to June stray birds may occur even in the south of India. The Sub-Himalayan tracts are those which tempt most birds to stay late, being well wooded and well watered.

During their stay fantail snipe are found all over the Empire, though their abundance varies in different localities, generally inversely with that of the pintail; in the southern and eastern provinces of India and in Burma, for instance, this species is the less common of the two, and often quite scarce, while it is the common species of the north-west part of the country. Snipe in distribution and the dates of it are, of course, somewhat affected by weather; heavy rains and prolonged cold weather will keep them longer in the north, delaying their spread to the

GALLINAGO GALLINULA $\frac{1}{2}$ GALLINAGO SCOLOPACINUS.

southern districts, and will likewise delay their final departure in the spring.

Although not actually flocking, as so many waders do, they are sociable to the extent that several may generally be found in one locality, and in the arrival and departure of the general body taking place simultaneously. Their chosen ground, as most people know, is swamp or marsh, wherever mud and low grassy cover is available; paddy-fields naturally appeal to them particularly. They resort to such places for food, which consists mostly, in the case of this species of snipe, of earthworms, although other small forms of animal life are also taken. The food has to be found by feeling, and the way in which a snipe's bill, by its flexibility, will open at the tip, the end only of the upper jaw being raised to nip the worm, is a wonderful adaptation to this mode of feeding, as is also the "overshot" structure, the upper bill ending in a sort of knob, back of which the lower fits at the tip, so as to penetrate with as little resistance as possible. This structure is seen more or less in all true snipe, and is generally a good distinction from the various sandpipers, whose bills are usually less adapted for experimental boring.

When they are not feeding, snipe like to be out of water, so during the heat of the day they are to be found on the nearest dry spot back of the mud on land, or even on water-weeds well out in a jheel. Being only small birds, also, they have no use for places where the water is more than an inch or so deep—too much water is just as bad as none at all from their point of view. Colonel Tickell sums up the situation by saying: "It is not easy to describe the ground this bird selects. In paddy fields, I found, where the stubble showed the mud freely—that is, was not too thick—and where *puddles* of water were interspersed, fringed with short, half-dry, curling grass and small weeds, there the snipe were sure to be if in the country; and note, if these puddles were coated over with a film of iridescent oily matter (the washings of an iron soil) the chances were greatly increased of a find."

The ground on which that celebrated snipe-shot, Mr. W. K. Dods, of Calcutta, made his record bag of 131 couple—259 of the present species, one pintail, and two jack-snipe—is described as "a large swamp tract of country covered with about the worst kind of 'punk' it has ever been my fate to shoot in, a black reeking mud composed entirely of decayed and decaying vegetable matter in which one frequently sank to one's thighs; growing in this ooze were dense clumps of hoogola reeds interspersed with fairly open glades, where birds could feed, and with other patches of thin null jungle in which snipe delight to rest during the day, secure from the too pressing attentions of the numerous hawks that infest these marshes." Snipe also seek cover in order to avoid the hot sun, for they are not birds of the light by choice, and at times even feed by night, besides migrating at that time. Their peculiar alarm-cry on rising—variously rendered as "scape," "psip," or "pénch"—is well known, as also the zigzagging in flight during the first few yards of their course. This style of flight is due to alarm, for a snipe can fly straight from the start if it wants to, and does so when going off undisturbed. The straight-away flight is swift, but it is generally agreed that snipe afford easier shots in India than in England, though there is some difference of opinion as to the advisability of firing at once when the birds rise, or letting them get their twistings over before "letting drive" at them; both methods have been practised by excellent shots.

Although their usual breeding haunts lie to the north and west of India, common snipe breed regularly in Kashmir, and very occasionally elsewhere with us. Mr. E. C. Stuart Baker took a nest himself and got the old bird as well, in the Santhal Pergunnas, and had another clutch brought him by his native collector. The birds when breeding produce the curious sound known often as "drumming," though it is better described as "bleating." There has been much discussion as to how it is produced, but the method seems to be now fully ascertained; the bird rises to a certain height in the air and swoops downwards, with the tail outspread and its two outer feathers standing away from and in front of the rest. It has been proved experimentally that these two feathers alone, properly manipulated, will produce the "bleat." In the most interesting experiment of all they were fastened to the notch-end of an arrow shot into the air, and the bleat came out as the dropping arrow reached the ground. Both sexes bleat, and they make this noise when alarmed as well as when courting. They also have a double note vocally produced, but not while drumming. Although in the breeding season snipe often perch on posts and trees, the nest is, as one would expect, on the ground, and is a very scanty affair; that Mr. Baker saw was composed of a fine, curly, brown grass. The eggs are peg-top-shaped, of an olive, drab, or brown colour, blotched with dark brown and lavender, and just over an inch long; four is the full set; the chicks run at once, and are mottled with light and dark brown and peppered with silver-white.

Snipe have many local names: *Tibud* or *Pan-lowa* among the Mahrattas; *Khada-kuchi*, in Bengal; *Kæswatuwa*, in Ceylon; *Mor-ulan* in the Tamil and *Mukupuredi* in the Telugu languages.

Pintail Snipe.

Gallinago stenura. *Myay-woot*, Burmese.

The pintail snipe is so like the common or fantail snipe in appearance, flight, and cry, that few people can distinguish it on the wing, though, as Mr. E. C. Stuart Baker tells us, a friend of his once won a wager with him by correctly referring to their species ten snipe, six fantail and four pintail, as fast as he shot them. When brought to book, however, they can be told apart with one's eyes shut. If one takes the bill of a snipe, of the ordinary type usually shot, in the thumb and fingers at the base, and feels it down to the tip, a distinct, though slight, thickening will be felt at the end in the case of a fantail, while in the pintail the calibre is practically the same throughout.

There is also a difference at the opposite end; on parting the soft feathers, or tail-coverts, which in snipe, as in most ducks, partly conceal the short tail, and counting the tail-feathers, there will be found on the fantail to be fourteen or sixteen in number, and all much alike and of ordinary shape, though the two outer are rather stiff and narrow—these are the "bleating" feathers, as remarked in the last article. In the pintail, there are ten ordinary feathers in the centre of the tail, while outside these are several pairs, up to eight, the number being variable, of curious short and very narrow feathers; these are those which give the bird its name, being little broader than a stout pin. Thus a fully developed tail in this species has twenty-six feathers. Specimens with only half-a-dozen pairs of pin-feathers in the tail are unusually large in body, and have particularly yellow legs; they weigh over five ounces, whereas ordinary pintail snipe average 3·91 ounces in the cock and 4·2 in the hen.

These big specimens very likely constitute a distinct local race, or sub-species, for as Mr. W. Val Weston, who first drew attention to them, says, they arrive at a different time from the ordinary pintail snipe, coming in with the fantails, which arrive in India later than the other species. Pintails may come in, though very rarely, in July, and regularly arrive in the beginning of August, but do not get down to Ceylon till October. On the whole they are more distributed towards the southern and eastern parts of our Empire than the fantails; I give the Burmese name because pintails are the snipe commonly got in Burma, for as a matter of fact natives seem never to distinguish between the two kinds, observant as many of them are. At the end of the year there are hardly any pintails in the north, but in March they are again the more abundant species in the north-east; and some may be found after the fantails have all gone north.

Thus, although in many places and at many times both kinds occur abundantly side by side, on the whole they tend to replace each other quite as much as to occur together. Another factor in their separation lies in a slight difference in their habits—as might be expected from the different form of the bill, which is less adapted for feeling in mud in the pintail—their food is rather different. Both eat worms, but while fantails chiefly consume water-snails and water-insects in addition, pintails consume land creatures in large quantities, land snails, caterpillars, and even beetles, grasshoppers, and flying ants. Such food is naturally sought on different ground, and so, though both are often found in the same places, pintail are often found feeding on grass land and in stubble fields, and will lie up for the day in jungle and dry grass.

F. Waller, Chromo Lith. 18, Hatton Garden, London.

$\frac{1}{2}$ GALLINAGO STENURA

In dry specimens a further difference in the beaks besides that of calibre is apparent; in the fantail the more abundant nerve-endings, drying up, give the end of the bill a much more pitted appearance than is seen in that of the other species, whose bill is less sensitive. It is probably on account of the less succulent food it consumes, as Mr. Baker suggests, that the pintail snipe is not *quite* so good on the table as the fantail, being often rather dryer in flesh.

Both species are liable to produce light, more or less albinistic varieties, of a fawn or creamy colour, and, far more rarely, very dark forms, of the type known in the common snipe in Europe as "Sabine's Snipe." Only one such dark blackish specimen of each kind has ever been recorded in India, and even the other variation, though commoner than varieties of birds usually are, is so rare that neither Mr. Baker nor Mr. Dods has ever shot one in all their long experience. The pintail is also particularly subject to minor variations in its plumage, the under-surface being often barred all over, while it has not the blank space in the barred wing lining which is noticeable in the fantail snipe.

Pintail snipe sometimes breed in the country; more than one set of eggs have been taken in Cachar, and in some years they seem to breed there quite frequently; the eggs are not certainly distinguishable from those of the common snipe, and the nest is similar.

The noise made in flight in the breeding season is, however, apparently characteristic, as one would expect from the very different tail-feathers; according to Prjevalsky, who studied this snipe in its breeding haunts on the Ussuri, "describing large circles above the spot where the female is sitting, it suddenly dashes downwards with great noise (which is most likely produced by the tail-feathers) like that made by our species and somewhat resembles the noise of a broken rocket." The vocal two-syllabled note, however, is probably much like that of the common snipe. The Yenisei River forms the western limit of the pintail snipe's northern breeding range, which is thus confined to Eastern Asia, and it winters in the East Indies as well as in India and Burma.

Swinhoe's Snipe.

Gallinago megala.

When the sportsman has grasped the difference between fantail and pintail snipes he can, if so disposed, find some mild additional interest in looking through the pintails in his bag to see if by any chance a specimen of Swinhoe's snipe has fallen to his lot; for this very rare species has only recently been added to our lists, and much resembles the pintail in most of its characters.

The distinctive point, as is so often the case with snipe, is to be looked for in the tail; in Swinhoe's snipe the tail-feathers are twenty in number, the six central ones being normal, while the rest, though decidedly narrow in comparison to them, are not so markedly so as to be strikingly noticeable and to be compared to pins. The tail is thus intermediate in type between a fantail snipe's and a pintail's.

It will be remembered that in the fantail all the tail-feathers look much alike, while in the pintail, which has about two dozen tail-feathers altogether, the side ones are very strikingly distinct, and though they are variable in number there are always at least eight normal ones in the centre.

Mr. Stuart Baker was the first to recognize this bird as an Indian species; he shot one himself at Dibrugarh in 1903, and had a skin sent him from the Shan States in December, 1908. That the birds should have been killed in these districts is natural enough, for the natural haunts of the species are Eastern Siberia and Mongolia, Japan, and China, whence in winter it goes to the Philippines, Borneo, and the Moluccas. One would, therefore, expect it would be more likely to turn up in Burma and Tenasserim, and it seems extremely likely that it has been overlooked, for though it is bigger than most pintails, there is nothing about it to catch the eye. In case it breeds anywhere on our eastern border hills, it may be mentioned that the eggs are said to be peculiarly shaped like a woodcock's, and pale cream or buff in ground colour with grey and brown spots.

½
GALLINAGO SOLITARIA

Eastern Solitary Snipe.

Gallinago solitaria.

There has been a good deal of confusion in the past between this bird and the wood-snipe, which is curious, because, although both snipe, and both big ones, they seem to lay themselves out, as it were, to be as different as possible from each other. The wood-snipe is as near a woodcock as it can be without actually being one; the solitary snipe is an intensified snipe in every way. It is the lightest in colour of all our snipes, as the wood-snipe is the darkest; it is a typical snipe in its flight, though naturally not so active as the ordinary birds, since it is a foot or more long, and weighs from five to eight ounces, as much as many woodcock we get here. Its shape is not in the least woodcocky, and its call is an aggravated snipe-call, "a harsh screeching" imitation of the note of the common snipe, says Hume, who notes that this bird goes off calling, while the wood-snipe is usually silent. In the hand the pure white belly of this bird, so different from the barred under-surface of the wood-snipe, is at once noticeable; its general appearance is that of the pintail rather than the fantail snipe, and it has several pairs of narrowed feathers in the tail, about the only point, apart from size, it has in common with the wood-snipe; although even here the colour of the feathers, white with dark bars, is quite different.

The solitary snipe is, it is true, a Himalayan bird like the wood-snipe, but it rarely penetrates into other parts of India, though it has been got as far away as Benares and the Wynaad, and is a regular breeder in the Chin and Shan Hills as well as in the Himalayas. In the winter it comes lower down, but very seldom strays from the bases of the hills. In summer it ranges up to 15,000 feet, and outside our limits breeds on mountain ranges in Central and Eastern Asia, and migrates south as far as Persia and Pekin. In spite of its title of solitary, it is not so much so as the wood-snipe, which is always alone, but may be found near one or two more of its kind as well as singly. It is nowhere numerous, though Hume estimates its numbers as at least ten times those of the wood-snipe; but it must be remembered that this bird is far less retiring, and is found in the low cover that satisfies ordinary snipes, and often haunts the margins of little streams in bare ravines where the cover is very scanty. Now and then, however, it may be found actually in forest, and Mr. Stuart Baker shot a breeding male in North Cachar in such a situation. Its bill is less sensitive and, therefore, less adapted for boring for worms even than that of the pintail, and its chief food appears to be small insects and tiny snails, and, although good eating, it is, in Hume's opinion, not equal to the rest of our snipes.

Its nuptial flight is much like that of the common snipe, but it descends from its pitch more slowly, and the sound it produces, though of the same general character, is recognizably different, as might have been expected from the different structure of the outer tail-feathers; it is harsher, and more of a buzz than a bleat. The nest is of a very slight character, and the eggs, four in number, are, according to Oates, easily distinguished from the eggs of all other snipes by their pinkish buff ground colour. They are clouded with dull purple and spotted boldly with dark brown, these spots tending to be elongated and to run in streaks. In the Himalayas breeding begins in May.

The Hindustani name, *Ban Chaha*, is the same as that applied to the wood-snipe, so that natives as well as Europeans seem to confuse these two very distinct birds. The Khasin name is *Simpoo*, the Assamese *Boner Kocha*, and the Cachari *Daodidap gophu*; the Nepalese *Bharka* simply means "Snipe" generally.

Western Solitary Snipe.

Gallinago major. "*Double Snipe.*"

This fine snipe, called great snipe by naturalists at home, is as big as our two large mountain species, but easily distinguishable from both by the outside tail-feathers being nearly all white and of normal width, this ordinary structure of the tail-feathers being a point to be noticed in young specimens, in which they are barred; four pairs of the feathers show this white, or white ground. Only one specimen shot in India has

GALLINAGO NEMORICOLA

GALLINAGO SOLITARIA

GALLINAGO SCOLOPACINUS

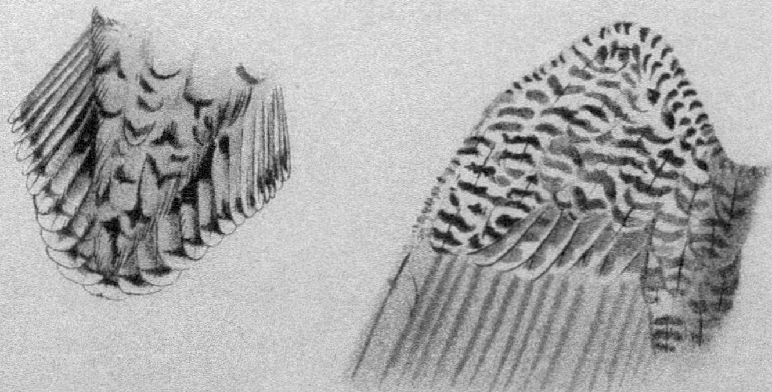

GALLINAGO STENURA

been preserved ; this was shot in October, 1910, near Bangalore, by Captain A. Boxwell, and was a quite young bird ; yet it weighed seven ounces. It rose without a cry, but with a pronounced flutter of the wings, from a patch of mud at the edge of a rice-field. It was not, however, the first specimen recorded, for another, an adult weighing over eight ounces, had been shot in September, 1899, near Madras, by Captain Donovan, who thought he had got the species, but lost the specimen through sending it to the Madras Museum for identification, when it went bad and was thrown away, its captor only getting the information that his prize was a "wood-snipe." I suppose, from the frequency with which this bird is brought out whenever there is a question of big snipe, that any bird of the snipe kind which is big gets put down as a wood-snipe because the name suggests the woodcock, known to be a big bird of the snipe kind ; for of course such mistakes ought never to be made. The moral is obvious ; in doubtful cases save the tail and eat your snipe. The great snipe breeds west of us, in Siberia and the north of Europe, and its usual winter haunts are the countries bordering on the Mediterranean, but it is also found as far east as Persia. Although in the structure of its bill—which is short for its size, not being so long as the fantail's—and its feeding-habits, it resembles the fantail snipe, as it does in its normally shaped tail ; it has a slow, straight, heavy flight. This sluggishness of movement may be the reason why it gets so fat ; the skin often breaks with the fall when the bird is shot. It is very nocturnal in its habits, and seldom moves by day. It is thus possible that it is often overlooked—indeed probable, as our only two recorded occurrences were in the South of India, where one certainly would not have expected it. Perhaps, like so many western birds, it may even breed in the Himalayas. It would be highly interesting if the double snipe ever does turn out to breed with us, because its breeding habits are unique in the group so far as is known. The birds appear to have no love-flights, but to carry out their courting exercises and vocal accompaniments on the ground ; this alone would be remarkable enough, but in addition they are social at this period, and hold tournaments after the manner of black-game and several other grouse, where the males show off, and fight when they meet each other. The fighting does not amount to much, being confined to feeble slapping with the wings, and not lasting long at a bout.

But the other performances of the males are curious; the bird runs about with puffed-out feathers and drooping wings, every now and then jumping on a tussock, snapping his bill and uttering a soft, nearly a warbling, note, audible for some distance and rendered as *bip, bip, bipbip, bipbiperere, biperere;* closer still various other sounds are audible, and the warbling amounts to a regular song. When singing the bird usually sits on a tussock, holding up his head till the snapping note is given out, and then depressing it, and erecting and spreading his tail till the white side-feathers show as two patches in the darkness; for although commenced at the oncoming of dusk, the "Spil" as it is called in Norway, is carried on all night. As the birds coming to it, in Professor Collett's standard account here indented on, are spoken of as so many pairs—usually eight to ten—it would seem they are mated already, and do not come to get mates, although the cocks are mentioned as running about as above described before the females.

The erection and spreading of the tail, by the way, is also noticeable in the woodcock under excitement.

The eggs of this bird are much larger than those of the common snipe, and vary much in the amount of their marking; the ground is olive-grey or stone buff, and the spots deep-brown and purplish-grey.

Jack-Snipe.

Gallinago gallinula. *Chota Chaha*, Hindustani.

The jack-snipe is a rather neglected bird in comparison with its relatives, perhaps on account of its small size; it is only about as big as a skylark, and its beak is less than two inches long, while the weight, although the bird is commonly very fat, does not exceed two and a half ounces, and may be an ounce less than this; there is no difference of size between the sexes.

More important points than size, however, characterize the

little jack ; curiously enough, he has, like the king of our snipes, the woodcock, only twelve tail-feathers, which, though the ordinary number for birds in general, is very short allowance for a snipe, as may be judged from what has been said about other species. All these tail-feathers, by the way, are pointed, but soft. More noticeable points, however, are the two cream-coloured stripes on the head instead of the three commonly found in snipes, and the marked sheen of green and purple on the upper plumage, which makes this little fellow the handsomest of the true snipes.

In flight the jack is notoriously distinct from snipes in general; he rises very straight, flickers like a butterfly, flies a very short distance and—another curious likeness to the woodcock—drops as if shot just when one does not expect it.

There is no doubt that, quite apart from the fact that the bird is often overlooked and sometimes despised for its smallness, it is not nearly so common or so regular in its visitations as the two stock snipes, pintail and fantail, nor is it so widely distributed as these are. Between 1894 and 1902, the years in which I lived in India, I only once found it common in the Calcutta market, although Hume says in his time it was brought in in thousands. Tickell and Mr. Baker, on the other hand, found it rare there, and, as we were all observing at different times, the inference is obvious that in some years, or periods of years, this bird visits us in greater numbers than in others, like so many other migratory species. It seems to be rare in the North-west, and has only once been shot in the Andamans.

As its breeding range, which extends all across the northern parts of the Old World, is to the north of that of our common snipes, it is not surprising to find that it does not breed here, but is merely a winter visitor, and has not been known to stay later than April, though it may arrive by the end of August.

When here it is very local, and a favourite haunt is closely adhered to, another tenant occupying it if the incumbent for the time being is killed, while it takes a lot of shooting at it to put a bird off its home. According to Hume, it is particularly fond of corners, whether formed by the embankment of a paddy-field or by natural cover, and it likes wet ground, but must have shelter as well, so that its attachment to certain spots is easily explained. Although, on its breeding-grounds, it somehow, in its nuptial flights, produces a great noise which is compared to the traditional "'ammer 'ammer, 'ammer, on the 'ard 'igh road" of a horse's hoofs, it does not give itself away by any note when rising, and its flight, if only by the mere fact that it is so different from that of other snipe, puzzles many people very much, though others say it is hard to hit. As Mr. Baker says: "Hume says that it is probably one of the easiest birds in the world to shoot if you reserve your fire till the proper moment, but I must personally confess that I have never yet quite made up my mind as to which this proper moment is!" But as to the jack's superior excellence on the table everybody seems agreed it is a case of "little and good" here, as it so often is in more important matters. Fortunately the bird, when flushed again and again, still adheres to its policy of lying close and not running, and so gives several chances. Dogs also mark it easily, on account of its unusually strong scent. As the food of this snipe includes grass-seeds and even a little grass itself, although chiefly consisting of the small forms of animal life eaten by snipe in general, it may fairly claim to be original in this respect also. It is worth mentioning that it is almost the champion egg layer of the bird world, for though the number is only the usual four, the eggs themselves are nearly as big as the common snipe's, and might easily be mistaken for them except for being less bulky.

Besides the native name quoted at the head of the article, others in use are *Obn* in Tamil and *Daodidap gajiba* in Cachari.

Wood-Snipe.

Gallinago nemoricola. *Ban-chaha*, Nepalese.

The wood-snipe is a very perfect connecting link between the typical snipes and those few members of the group which are dignified by the title of woodcocks; in style of flight, in the dark colour of its plumage, and especially in the dark transverse bars all over the lower parts of the body, it is a true woodcock,

GALLINAGO NEMORICOLA

while it is also bigger than most snipes, being a thick-set bird weighing about five ounces and sometimes more, and measuring a foot in length. On the other hand, it shows snipe points in the bare hocks and the longitudinal dark markings on the head—those of all woodcocks being transverse—and in having several pairs of narrow feathers at the sides of the tail. The brown colour of these, by the way, is one of the distinctions between the wood-snipe and another big snipe often confounded with it, the Eastern solitary snipe, which also has narrowed lateral tail-feathers, which in its case are white with black bars.

The plain dark pinion-quills, so different from those of the woodcock with their chestnut chequering, are at first sight a distinct point of essential snipiness; but the American woodcock (*Philohela minor*) which everyone would call a typical woodcock, has also plain quills, so that, though useful to distinguish the wood-snipe from a small woodcock, they do not count either way in estimating its affinities.

Even the haunts of the bird show its intermediate nature; it frequents not so much woodland itself, but thick high grass cover at the edge of woods, wherever the ground is swampy or contains small pools; the grass it affects is such as would be far too high and tangled for ordinary snipe, and it lies close and does not go far when flushed. Although, as in the case of the woodcock, stray specimens may occur away from the hills, this species again resembles the head of the clan in being essentially a mountain bird, breeding in the Himalayas, and visiting the hills of southern India during the winter. It extends to Manipur and even Tenasserim, but has not yet turned up anywhere outside our Empire.

Owing to its partiality for the unhealthy swamps of the Tarai, it is a very little known bird, and besides appears to be really scarce, although its retiring disposition and its habits, which seem quite as nocturnal as the woodcock's, no doubt cause it often to be overlooked; in many places it can only be shot from an elephant's back, and people hunting in this way are generally after something better than snipe!

When rising, if it calls at all—it is generally silent, woodcock fashion—the note is a double croak, rendered as "tok-tok."

The fact that it has modified feathers in the tail no doubt indicates that it performs love-flights and makes tail-music—a snipe point—but as a matter of fact there seems to be no record of this. The breeding-zone of the wood-snipe is lower in the Himalayas than that of the woodcock, for though it may be found breeding up to 12,000 feet, and therefore in the woodcock's territory, it also breeds as low as 4,000, at which elevation Mr. Stuart Baker took a nest near Shillong, while in Manipur it is suspected of breeding at 2,000 feet, and Mr. Baker rather thinks it is resident in the Himalayan dooars.

The egg he got from this nest seems to be the only authentic one in existence and, unfortunately, it happened to be a dwarf one; the others in the clutch were smashed by the struggles of the parent, which was snared on her nest by his collector, but as the man said they were all alike, it is enough to indicate the colour, which, says Mr. Baker, is like that of many common snipe's eggs, but unusually brown; the shape also is quite ordinary. The food of the wood-snipe consists rather of insects than of worms, and curiously enough of small black seeds also; it is itself a particularly good bird for the table, if not of much importance as an object of sport.

Woodcock.

Scolopax rusticola. *Simtitar*, Hindustani.

Plenty of good sportsmen have never shot or even seen a woodcock, just as Tickell says was his case, his first impressions of the bird, in Nepal, being as follows: "Imagining from the general resemblance of the two birds that a woodcock must fly like a snipe, I was much taken aback, when hailed to 'look out,' at perceiving what appeared like a large bat coming with a wavering, flagging flight along the little lane-like opening of the wood where I was posted; but in an instant, ere I had made up my mind to fire, the apparition made a dart to one side, topped the bordering thicket, and seemed to fall like a stone into the covert beyond."

When this queerly behaving bird is brought to bag, it is seen

SCALOPAX ½ RUSTICOLA

to be indeed like a snipe in the very long straight bill, overshot at the tip, and the peculiarly far-back position of the eyes, seen to perfection in this bird, in which they are very large; but the bird is as big as a good pigeon, and pigeon-like in its shortness of leg and absence of bare skin above the hock, while the plumpness and short tail rather suggest a duck, and the broad wings sufficiently explain the un-snipe-like flight. The mottled brown plumage, though very characteristic when one knows it, has nothing to catch the eye at first sight, except perhaps the three broad black bands across the back of the head, which are the peculiar coat-of-arms of all true woodcock.

It is worth while going into these details about so well known a bird, for Blanford says black-tailed godwits were sold in the Calcutta Bazaar as woodcock, and though this was not so in my time, it shows that many people did not know this valued sporting bird and table delicacy by sight; for though this godwit is much about the same size in body as a woodcock, and has a very similar bill, it has a long neck and typical waders' legs, and a quite different plumage from that above described. I only once saw a woodcock in the Calcutta Bazaar, and that looked as if it had not been killed recently; and, as a matter of fact, a woodcock found anywhere away from the hills in India may be put down as "lost or strayed," though in Burma they come down to the plains much more than in India. In the hill-regions they occur as far south as Ceylon and Tenassarim, and woodcock-shooting is quite an established sport in the Nilgiris. Mr. Stewart Baker sums the matter of the woodcock's Indian distribution up by saying that "anywhere between November 1 and March 1 on hills over 4,000 feet elevation one should be able to find woodcock if sufficient time and trouble is given to the search, and there are suitable places for the birds to lie up in." Such suitable places are where woodland cover is near swampy spots in which the birds can feed, and these spots in hills are naturally usually by streams, which has given rise to the idea that the woodcock especially requires running water. It is, everywhere except in the Himalayas, a cold-weather bird, but in that range at heights of over ten thousand feet, it is a well-known breeder, as it is all across Europe and Northern Asia.

What with its goodness for food, and its unmistakeable appearance—testified to by its many native names, *Simkukra* in Kumaon and Nepal; *Bumpal* or *Dhabha* in Chitral; *Chinjarole* in Chamba; *Kangtruk* in Manipur; *Daodidap gadeba*, in Cachar; *Simpso Khlan*, among the Khasis; *Gherak*, in Drosh; and *Chustruck* in Gilgit—this bird always attracts attention everywhere, in India as well as in Europe, and the uncertainty of finding it adds to the value put on it. And, as it rises without warning, and not unless it can help it, and has perforce to dodge if its line of flight takes it through trees, shooting it is generally a triumph. Hume, however, regarded it in India as a sluggish flier and an easy shot, only worth firing at for its goodness on the table, but this experience is by no means universal, though, like snipe, the bird tends to be tamer and easier to hit than in England.

The fact that the woodcock digs his bill in the ground, and swallows his worms something like a duck, not tossing them down his throat by an up-jerk of the bill, as a crane or stork would do, has no doubt given rise to the idea that he lives by suction, this notion being aided by his very rapid digestion. He is really a far greater glutton than the much maligned vulture, which only sees food, probably, about once a week, and then has to scramble for it, while the woodcock, essentially a hermit, "does himself well" every night of his life, and can put away an incredible number of worms—a tame bird will eat a cupful at a sitting. Insects, both in the larval and adult states, are also eaten, and even frog-spawn does not come amiss; the bird's nights must be pretty fully occupied in getting enough food, for it is strictly a night-bird, and seldom moves by day.

At evening and grey dawn too, the bird's courting manœuvres are carried on; his love-sport resembles that of the snipe in being aerial, but he does not drum, but flies to and fro, in crescent paths of fifty to two hundred yards, uttering alternate croaks and squeaks, and getting lower at each turn, though at first above the trees; this is called "rôding" on the continent. At such times the birds are easily shot, but this is mere poaching, as both parents are needed to attend on the young. These are beautiful little creatures with comparatively short bills and velvety tortoise-shell down, and when the family has to be moved in order to avoid a foe, or even to seek food, the old birds, or at any rate the hen, actually carries them, a habit very rare in birds. It was long a puzzle how this was done, and one picture even depicts the old bird with the baby riding pick-a-back, but as a matter of fact observers both in Europe and in India have established that they are really held between the legs; and they are carried thus even when half-grown. Four is the number of the family, the eggs producing them are laid on a mere bed of the dead leaves found in the bird's usual haunts; they are of some shade of drab or buff in ground-colour, and rather sparingly marked with brown and grey spots of various tones and distribution; they are about an inch and three quarters in length. Indian eggs are not smaller than European, as Mr. Baker points out, and he also shows that the idea that the birds themselves are smaller in India is due to the fact that it is the immature birds which are shot, these being those which migrate south; but no Indian specimen has yet been shot weighing as much as a pound, which they often do in Europe. The idea that the plumage shows any difference with age has also been exploded, but some individual birds are much greyer than others.

Painted Snipe.

* *Rostratula capensis.* *Kone*, Kols of Singbhoom.

When out snipe-shooting, if a big specimen gets up and flies straight off with an indolent fluttering flight and legs dangling at first—moorhen-fashion, in fact—it may be known at once as a painted snipe. Or if hit without the flight being noticed, and not killed, the swearing hiss when approached, and pitiful attempts at menace by the spreading of its spotted wings, again give away the painted fraud; for this bird, the most beautiful of all its tribe, although belonging to the same family as the true snipe, cannot rightly be referred to the same section,

---

* *Rhynchæa bengalensis* on plate.

but if it must be given a place, has to take one among the humble sandpipers, though among these it stands quite alone.

Its bill, curved downwards slightly towards the end, is as distinctive as its fine plumage, of which the most striking points are the blue-grey quills and tail, spotted boldly with buff, a coloration unique in the bird world.

The sexes differ much, though both have, in addition to the same peculiar coloration of the quills, the snow-white abdomen; the hen, which is larger, is also much the handsomer, her back being dark glossy green, with a streak of white on each shoulder, and her neck dark chestnut.

The cock's back is mottled with buff on a much duller green, and he has a buff ring round the eye where the hen has a pure white one; and, most noticeable of all, his neck is only drab, not chestnut. His daughters, as is usually the case where the hen bird wears, if not the breeches, at any rate the fine feathers, have the masculine plumage as their first dress. Young chicks differ noticeably from those of the true snipe, being buff with a few distinct longitudinal markings of black, a style of coloration more reminiscent of sandpipers. Although the "painter"—to use a slangy name, but appropriate as being non-committal with regard to the owner's relationships—is only about as long as ordinary snipe, it stands higher on the legs and is much more strongly built. It is, in fact, a broad-shouldered, full-chested bird, and in this respect differs much from the slab-sided rails, which it otherwise much resembles in some points, notably its slovenly flight and habit of slinking along head down when alarmed into cover, and running along therein in preference to rising and showing sport. It will also swim voluntarily, as rails so usually do.

It is not, indeed, generally regarded as a sporting bird, at any rate when genuine snipe are about to shoot at; for in addition to being a skulker and a slack flier, it is no particular delicacy, though not unpalatable in default of more savoury game; I should call it about as good as an ordinary pigeon.

It is a resident, or at least does not migrate more than is necessary for any marsh-bird when its haunts are dried up, and it is found practically everywhere in our limits except in

$\frac{1}{2}$

RHYNCHEA BENGALENSIS

the hills, so long as cover and water are available. Muddy ground with plenty of shelter suits it especially, but it does not frequent paddy-fields much. In some places it appears decidedly sociable, and flocks of up to twenty birds may be met with.

In the breeding-season it would appear, from information given to Mr. Stuart Baker by Cachari shikaries, that the hens (which they mistake for cocks) fight vigorously for their mates, just as hen hemipodes do; and there seems to be now no doubt that the cock painter does the sitting and rearing. Several of this sex have been caught on the nests, but never their ladies, who, gay in more senses than one, are suspected of roving off in search of a fresh *liaison* when they have got one husband comfortably settled on a quadruplet of eggs. The eggs are not generally so peg-top-shaped as true snipe's eggs, and are often of the usual oval, while they usually vary between the two types, with an inclination to the former. They are very handsomely coloured, the ground being of a buffy yellow, shaded with green or grey or some other tint, and the spots are large and nearly black, with a few markings of pale brown as well.

The nest is better constructed than that of the true snipes, at any rate in many cases; it is made of grass, weeds, &c., and is sometimes quite hollowed out if in a natural hollow. It is occasionally placed, not on the actual ground, but on thick grass or other marsh vegetation a little above it, and though generally well concealed, is by no means always so. This is one of the birds whose breeding arrangements evidently depend entirely on the food-supply; it may be found nesting all the year round in some part of its range, and is even suspected of breeding twice a year when its lines are cast in particularly pleasant places. This is not surprising, when we consider its free-and-easy matrimonial ideas, which relieve the female of all work in rearing, and its omnivorous nature, which admits of its feeding freely on paddy and other seeds, and paddy leaves, as well as on insects, snails, and worms; for which, by the way, it does not bore, at any rate in captivity.

It seems to be a nocturnal bird, but Mr. Stuart Baker has found it feeding in open ploughed fields by day except during the hot hours; but this was at a time when tiny crickets were abundant, and I rather think that, as the gentleman alluded to suggests, the birds which were feeding freely on them were acting abnormally in consequence of this abundance, as birds often will, and this may have caused them to become diurnal as well as indifferent to cover.

The note of this bird is said to be like the noise produced by blowing into a phial, expressed by the native name *kone*. Wood Mason says the male squeaks in answer to the "low, regular, hoarse, but rich purr" of the hen; but Hume, who considers the note to be the breeding call, heard no other, and personally I have only heard what I call "swearing" from captive birds, and noticed no difference. Yet there ought to be some, as the female, according to Wood Mason, has a longer looped windpipe, a peculiarity which is exaggerated in the Australian painted snipe, which otherwise differs little from ours. Our bird has an enormous range, being found nearly all over Africa and southern and eastern Asia; in fact, it is one of the most widely distributed of the usually non-migratory birds.

It is curious that so remarkable and easily recognized a bird should be so little distinguished by native names; but it is called *Ohari* in Nepal, and *Mailulan* by Tamils, while in Ceylon the Cingalese distinguish it appropriately as *Raja Kæswatuwa*, the king snipe.

Next to the snipes the godwits may be considered.

Black-tailed Godwit.

*Limosa belgica.** *Gudera*, Hindustani.

With the build of a miniature stork of a pigeon's size, the legs, neck and bill all being long, and with the contrast between its short, pied tail, black at the tip and white at the base, with its drab plumage as it rises, this godwit is a conspicuous bird, and ought to be well known to sportsmen. The size mentioned above is only approximate, for this is one of

* *ægocephala* on plate.

LIMOSA ÆGOCEPALA
SUMMER & WINTER PLUMAGE

the most variable birds in dimensions that exist, if indeed it does not surpass any in this respect. Weights run from less than half a pound in the case of the smallest males to within an ounce of a pound for big females, the birds of this sex running far larger than their mates, though there are plenty of big males larger than many females.

The bill of the godwit is not sharp like a stork's, but blunt and overshot at the tip, much like a woodcock's, though not sensitive; one would never expect a bird with this type of beak to eat grain, yet this species is quite as fond of grain as any partridge or duck, and feeds by preference on rice whenever it gets the chance, as well as on millet and grass seed. It does, however, also devour worms, grubs, shrimps and shell-fish, the ordinary sort of food one would expect a long-billed wader to take, in fact. Whatever the food is, the bird's flavour is uncommonly good, and Hume considered rice-fattened plump specimens as good as the woodcock or jack-snipe, though with a different flavour.

It is fortunate therefore that these birds, though often seen in ones or twos, are also commonly found in large flocks, and are widely though locally distributed during the cold weather, the only time when they are to be found in India. They are not to be expected in any numbers before the end of October, but few stay on till the beginning of April. During their season they may be met with here and there all over the Empire, except in the Andamans and Nicobars, but are rare in all the southern provinces, and not common east of Bengal. The commoner they are, the larger the flocks met with, and the easier are the birds to get near. They frequent both the coast and inland waters, keeping away from cover, and wading, or resting on one leg, in the shallow margins of swamps and jheels. They feed, being so fond of rice, in the rice-fields by preference, and will do so either by day or night, according to the amount of disturbance they have been meeting with.

Their resting, as opposed to feeding, places are the shallows above mentioned, but here also they pick up a good deal of food, both on land and in the water. Their flight is straight, fast and high, though, like so many excellent fliers, they rise heavily. If anywhere near, the white bar on the wing as well as the white base of the tail is conspicuous, and their long, straight bills are also characteristic. They are very silent in India, but may have a whistled alarm-call as they rise.

They breed in the temperate zone of the Northern Hemisphere, and at the breeding-time have the plumage mainly chestnut, not drab, with some black mottling on the back, and barring below; most birds before leaving us show a good deal of this plumage coming on.

The numerous native names of this bird attest its familiarity: The Bengali one is *Jaurali,* the Nepalese *Malgujha,* and the Telugu *Tondu ulanka;* while, besides, *Gudera, Gairiya, Jangral,* and *Khag* are Hindustani names.

Bar-tailed Godwit.

Limosa lapponica.

The bar-tailed godwit is a sort of poor relation of the better known black-tailed species, smaller and less strikingly coloured, the tail lacking the bold contrast of solid black and white in two sections, but simply marked with many transverse bands of brown and white, while the body plumage is longitudinally marked with dark stripes.

Although the dimensions of this bird average smaller than those of the last, there is some overlapping, the smallest being barely lighter than small examples of the black-tailed, though the largest do not reach twelve ounces; these, as in the common species, are hens, this sex averaging bigger in the bar-tailed godwit also.

The bar-tailed godwit, although more familiar at home nowadays than the black-tailed, is little known in India; it is, however, common along the south coast, where specimens have been got in Kurrachee harbour, one as early as September 29. The bird is only a winter visitant, and the latest was got on March 23. At Kurrachee, according to Hume, these godwits haunt the extensive mud-banks, mixed up with other waders, but flying off in flocks when alarmed. These flocks did not exceed twenty birds, and however many were

2/5
LIMOSA RUFA
SUMMER & WINTER PLUMAGE

together when feeding, they did not go off in one big flock, but split up into smaller parties and each took its own line, flying less swiftly than the black-tailed species, though rising quicker. They were so wary that he only got six specimens, though as many as a hundred birds might be seen on one bank at a time, and they were even more silent than the other kind, very occasionally uttering their low pipe; their food had consisted of small sea animals, and they themselves had the peculiar flavour which Hume calls "froggy," reminding him of eels from muddy water; this is curious, as even when near the sea the black-tailed godwit retains its excellent flavour. Like that bird, the bar-tailed godwit puts on a chestnut plumage in the breeding-season, though retaining its characteristic differences, and it also breeds all along the Northern Hemisphere. In winter a race of it even reaches New Zealand and is a favourite object of sport with gunners under its Maori name of *Kuaka*, or the very misleading English one of "curlew."

Snipe-billed Godwit.

* *Macrorhamphus semipalmatus.*

This very rare bird has exactly the bill of the most typical snipes, overshot, broader at the tip than the middle, and so soft and full of nerve-endings at the tip that it becomes pitted when drying after death; it can be distinguished from all true snipe and their painted imitator, however, by having the toes webbed at the base, whereas the toes of snipe are free to the very root.

In this basal webbing of the toes the snipe-billed godwit shows its relationship to the godwits proper, although in these the web is less developed, and its plumage is also a godwit's, not a snipe's, being in winter variegated with drab and whitish, without the rich dark tints and creamy head- and back-stripes so usual in snipe, and in summer a bricky red with dark markings on the back. It is nearly of the size of small male specimens of the bar-tailed godwit, being rather over a foot long, and it

* *Pseudoscolopax* on plate.

resembles that bird in the barred colouring of its tail, so that it might easily be confused with it, were it not for the fact that the plumage does not show the distinct dark streaking of the god-wit's upper parts, and that the bill is truly snipe-like and not tapering, but bulging at the end.

Extremely little is known about this bird, which has not been found commonly anywhere; it is supposed to breed in Siberia, and a few specimens have been got in north-east Asia. A few also have been obtained in our Indian Empire in the cold weather, at intervals of many years.

Jerdon seems to have got the first recorded specimen in the Madras Market in 1844; and since then it has been obtained in that of Calcutta by Blyth in 1847, and Hume in 1878; and near this time Oates shot a pair in Lower Pegu. This bird has also been killed in Assam, and is known to occur in China and as far to the south-east as Borneo. But hardly anything is known about it, which is the more to be wondered at, as its only near relative, the so-called red-breasted snipe (*Macrorhamphus griseus*) of America, is well known to shore-gunners there, and has even strayed to the British Islands. Among other birds of this group only greenshanks and golden plover can be noted.

Greenshank.

Totanus glottis. *Timtimma*, Hindustani.

The greenshank is not only the biggest, but much the best for the table, of the various sandpipers or snippets, most of which are contemptuously passed over by sportsmen as birds of no account; but as it is really good, and easily recognizable, it is worth mentioning here, especially as the front figure in Hume's plate of Armstrong's yellowshanks might easily be taken for it.

It is a most graceful, elegant bird, with a straight, slender, pointed bill, and greyish plumage with conspicuous white under-parts and rump, which last, together with the nearly white tail, is noticeable as it gets up with its characteristic shrill cry, imitated by its native name in Hindustani; the Bengali name is *Gotra.* The legs are green and the bill black at the tip

½
PSEUDOSCOLOPAX SEMIPALMATUS.

and clear blue-grey at the root; the whole length about a foot and a quarter, while the body is as big as a small partridge's. The greenshank is a winter visitor, staying from September to April; it is distributed all across the northern parts of the Old World in summer, and visits Australia as well as India and China in winter. In summer plumage the fore and upper parts are much streaked with black. The marsh sandpiper (*Totanus stagnatilis*) is like this bird on a small scale.

Armstrong's Yellowshanks.

*Totanus guttifer.**

Distinguished from the greenshank by its darker tail, rather smaller size, and proportionately shorter legs, which are yellower, this rare bird is probably often passed over; it is an East Asiatic bird, breeding in the north. In winter it has been got in Hainan, and Hume got it in the Calcutta Bazaar, while Armstrong also obtained it at the mouth of the Rangoon River; but hardly anything is known about it. In summer plumage, the back is nearly black.

To distinguish this bird certainly, attention should be paid to the length of the shank from hock to toes, which is less than two inches and less than the length of the bill; whereas in the greenshank the shank similarly measured is 2½ inches, and the bill just equals this.

Eastern Golden Plover.

Charadrius fulvus. *Battan*, Hindustani.

The Eastern golden plover, whose yellow-speckled plumage and whistling call make it so distinct from other game-birds of the marshes, is a very well known bird in the East, going about in flocks, and being found not only in India, Burma and Ceylon, but even being common in the Andamans, Nicobars, and

* *haughtoni* on plate.

Laccadives. It becomes, however, rare in Sind, and is not found in the hills nor in jungly districts; it likes flat swampy districts near the coasts and large rivers. It is an animal feeder, devouring insects and worms, and is good eating, though said to be not equal to the European golden plover.

Golden plover come in in September, and, though Jerdon's statement that they breed in India is questioned nowadays, at any rate they may be found in mid-May. At this time they have black under-parts, making a striking and beautiful contrast with their yellow-marked backs, and outlined by a pure white border.

This plover breeds in the northern parts of the Old World from the Yenisei eastward, and also in North America; but the American race is larger than the Eastern. It visits the Malay Archipelago, Australia, South America, and even the Sandwich Islands, in winter, but is very rare to the west of our limits even at that time. It should be noticed that both this and the following species, like so many plovers, have no hinder toes.

European Golden Plover.

Charadrius pluvialis.

The European golden plover, although a rarity in India, is probably often passed over, for it is so like the Eastern species that it is not likely to be distinguished except in the hand; the difference is then readily perceptible, the under-parts of the wings being white, whereas they are greyish-brown in the Indian golden plover. The European golden plover is also a heavier bird than the Eastern, although its legs are not any longer; American specimens, however, nearly equal it in size. The young of the grey plover (*Squatarola helvetica*), which is found in India in winter as well as almost everywhere else, might easily be mistaken for very large golden plovers, being spotted with yellow, but they have a small hind toe.

F. Wailer Chromo-lith. 18 Hatton Garden London
½
TOTANUS HAUGHTONI

Indian Water-Rail.

Rallus indicus.

This species, with its long bill and toes, short wings, and flat-sided body, is an excellent type of the rails in general; although the length of the beak and toes is variable, the general appearance of a rail is unmistakeable in the hand, and even when at large the slinking gait, flicking up of the tail, aversion to leave cover, and heavy fluttering flight with legs drooping at first, mark off these otherwise insignificant birds at once.

The Indian water-rail is about as big as a snipe, but its bill is shorter and stouter, though still noticeably long; as in all rails, the beak is hard and strong, not soft as in snipes. The plumage is very unpretentious, but yet recognizable, with the black striping on the brown upper-parts, grey face and breast, and black-and-white zebra-barring of the sides. The only bit of bright colour is the red at the root of the lower jaw, and even this is replaced by a yellowish tint in young birds.

This rail is simply a local race of the well-known water-rail at home (*Rallus aquaticus*), and this Western typical form has sometimes been got in the Himalayas in winter. The only differences between the two are, that in the Indian bird there is a dark streak running back from the eye, and a wash of brown on the breast, these parts being pure grey in the Western water-rail. The Indian bird itself is not a resident in India, its home being in north-eastern Asia; but it is a fairly common winter visitor to the northern provinces of India, and ranges as far as Arrakan. It is found in grass or rush cover on wet ground, and is difficult to put up; its food consists of small snails, insects, worms, &c., but it also takes some vegetable food, seeds, leaves, and bulbs. These long-billed rails may be more insectivorous than the shorter-billed kinds, but rails as a family are quite as omnivorous as the game birds and ducks. The breeding call-note of the European water-rail is a groaning sound, called "sharming"; the Indian bird croaks like a frog, but probably has a similar note in the breeding season.

Blue-breasted Banded Rail.

Hypotænidia striata. *Kana-koli,* Tamil.

This very pretty bird is also about the size of a snipe, with a distinctly long bill; the face and under-parts are grey, much as in the Indian water-rail, and the flanks similarly barred with black-and-white, but there is a distinctive point in the cap of chestnut covering the head and running down the neck, and in the broken white pencilling on the brown back. This white marking is wanting in young birds, which also have the cap less richly tinted, but it soon begins to develop. Hens are less richly coloured than cocks.

Although the bill of this bird is long, it is not so much so as in the Indian water-rail, and is thicker for its length. It is a widely distributed bird in our Empire, except in the North-west, but in the Andamans is represented by a larger race—the so-called Andaman banded rail (*Hypotænidia obscurior*), which is much darker all over, the cap being rather maroon than chestnut, the breast slaty, and the back blacker.

The banded rail is not quite such a skulker as the water-rail, though it frequents the same sort of grass and mud cover on wet ground, and feeds in a similar way; now and then four or five birds together may be seen out feeding on turf or grassy banks near the rice fields or wet thickets in the early morning, but commonly they go singly or in pairs. Like button-quail, they will rise to a dog readily enough the first time, but will risk capture rather than get up again; and they do not fly many yards in any case. They can swim if put to it, but are not water-birds in the sense that moorhens and some of the crakes are.

This species is not apparently migratory, though a wide-ranging bird, and found throughout south-east Asia to Celebes. It nests at the water's edge in grass, rice, or similar cover, making a pile of rushes and grass, and laying about half a dozen white or pink eggs with reddish and mauve markings of various sizes, chiefly towards the large end. They may be found as early as May, or as late as October. This bird is called *Kana-koli* in Telugu; in Burmese *Yay-gyet.*

RALLUS ½ INDICUS.

Banded Crake.

*Rallina superciliaris.**

The banded crake, although barred with black-and-white below like the long-billed rails, has a shorter bill and toes than these birds, though the beak has not the almost fowl-like form found in some of the short-billed crakes and the moorhens; it is rather over an inch long, the bird being about as large as a snipe.

The general colour above is chestnut in front, and greenish-brown on the back; but only in adult males is the chestnut developed all over the head and neck; young birds are brown even here, and hens, except perhaps old ones, have the crown and back of the neck brown. The legs are grey, and the bill brown and green.

The banded crake is seldom found outside Ceylon; but it is believed to have been found breeding about Karwar and Khandalla, and specimens have been got in many other localities in the Empire, from Oudh to Singapore, so that it might be expected to occur anywhere. Even in Ceylon it is only to be found from October to February, and does not breed in the island so far as is known. While there it inhabits the hills, and though often found in the usual haunts of rails, in cover by streams and paddy fields, it is not confined to wet ground. The setting in of the north wind is the signal for its arrival in Ceylon, where it first appears on the west coast. The incoming birds behave as if they had made a long journey, being very tired and taking refuge in all sorts of queer places. Layard says that he found one in the well of his carriage, one in his gig-apron, and another in a shoe under his bed! The bird, in fact, seems to have quite a mania for coming indoors.

When flushed this bird quite commonly takes to a tree, and the nests attributed to it, and found in the mainland localities above mentioned during the monsoon, were placed above ground on bamboos, tangled herbage, bushes, or stumps. The eggs were creamy-white and unspotted, thus being abnormal for the family, since spotted eggs are general among rails.

---

* *euryzonoides* on plate.

Malayan Banded Crake.

**Rallina fasciata.*

The Malayan banded crake can easily be distinguished from the rest of our rails with zebra-barred flanks by its bright red legs; other rails with legs so coloured have not the black-and-white side-stripes. The present species is of a pretty uniform reddish-brown in front and above, and is a smaller bird than the banded crake, only measuring about nine inches.

As its name implies, it is a Malayan bird, but extends to Karennee, and in the other direction to Celebes. It is found in Tenasserim, but not commonly, and likes the vicinity of cultivation if this provides suitable cover; rice fields are favourite haunts when surrounded by scrub, but not when situated among dry forest without undergrowth.

Andamanese Banded Crake.

Rallina canningi.

In many cases the Andamanese representatives of Continental, Indian and Burmese birds can hardly be called more than local races, but some, like the present bird, are very distinct. Although its black-and-white barred sides and brown general colour show its relationships at once, it is far bigger than our other two banded crakes, being about as large as the common grey partridge. It stands high on its legs, and has the bill and toes comparatively short, and the tail over three inches long, so that its proportions are less typically rail-like than usual. The brown of its body is deep rich mahogany-red, and the black-and-white bars of the sides and under-parts are very clear and striking; moreover, although the feet are only olive-green, the bill is of the clearest apple-green, and looks like jade. Thus, even if occurring on the mainland, this handsome crake would be noticeable not only among other members of the rail family, but among birds in general.

In its native islands the crake frequents forests, and is not

* *Porzana* on plate. † *Euryzona* on plate.

½

HYPOTŒNIDIA OBSCURIORIA — HYPOTŒNIDIA STRIATUS

to be seen in the day-time unless driven out of its cover, when it flies slowly and heavily. Like rails in general, it keeps near water, or at least on moist ground. It feeds on insects, and can be taken in snares baited with shrimps. The nest is on the ground, and the eggs are spotted with purple and maroon markings on a ground of white or stone-colour with a pinkish tinge.

Spotted Crake.

Porzana maruetta. *Gurguri khairi*, Bengali.

Known in Telugu as *Venna mudikoli,* the possession of even two native names shows that this pretty bird is fairly well known, though only a winter visitor. It is short-billed but long-toed, and about the size of a snipe ; the speckling of white all over the plumage is characteristic, the ground-colour of this being of a common rail pattern, streaky-brown above and grey below in adults, though young birds have a brown breast. The sides are vertically barred, but the dark interspaces between the white bars are grey, not black ; the yellow bill is also a noticeable point.

The spotted crake arrives in India in September, and leaves about April ; it mostly visits northern India, though in Jerdon's time it seems to have been more generally distributed ; to the coast it extends as far as Arrakan. It particularly frequents rice-fields, rushes and sedge, and has a great objection to exposing itself in the open, while if it is forced up, it drops after a flight of about a score of yards, and declines to appear again. It is worth shooting if come across, as it is good eating, according to Jerdon. It feeds on water-insects, snails, as well as on seeds and herbage.

Generally it is found singly, and in any case not more than a pair seem to keep about the same spot. The call-note, mostly heard at night, is a clear loud "*kweet*," according to Dresser. This is a widely distributed bird, breeding from our own islands to Central Asia, and, in spite of its great reluctance to fly in the ordinary way, appears to cross the high Karakorum range in its southward migration to the Indian Empire.

Little Crake.

Porzana parva.

The little crake is a short-billed and long-toed little bird about as big as a lark, with streaky brown rather lark-like plumage, slightly variegated with white above, and grey below on the old cock, while the under-parts of hens and young birds are buff, shading into brown behind, where the plumage is diversified with white cross-bars in all. Young birds are more freely barred, and show more white on the upper-parts.

The little crake is the smallest of our rails except the next species, but is a comparatively bold bird, being found running on water-lily leaves and swimming in the water between them ; it also appears to dive quite freely, and is altogether more of a water-bird and less of a swamp-runner than most rails. It is also a pretty good flyer as rails go, and is only a winter visitor to India, and then only to the extreme North-west, its chosen haunts being the broads in Sind, where it feeds on water-insects.

This, like the spotted crake, is a well-known bird in Europe, and does not extend farther than Central Asia to the eastward.

Eastern Baillon's Crake.

* *Porzana pusilla.* *Jhilli*, Nepalese.

This little bird is even smaller than the little crake, but closely resembles it, having a black-streaked, white-splashed upper-surface and under-parts grey in front and with white cross-bars behind ; but in the present bird the cock and hen are alike, and it is only the young which differ in having the breast and throat buffish instead of grey. But the easiest way to distinguish these two tiny crakes, or pigmy moorhens as they might be called, from their habits, is to remember that in the Eastern Baillon's crake the first wing-quill has a white edge, whereas in the little crake this is not the case.

This Eastern race of Baillon's crake of Europe, the original *Porzana bailloni*, has a dark-brown streak along the face which is

---

* *Crex bailloni* on plate.

PORZANA ½ EURYZONOIDES.

wanting in the Western form ; it is generally distributed in India and Burma, and is generally resident, though a good many come in in the cold weather from countries to the northward. It reaches not only Ceylon, but the Andamans, and breeds as far south as Tavoy.

In the plains it may be found nesting up to September, but though nesting begins about the same time in the Himalayas —in June—it does not go on so late there. Wild rice, or rice cultivation, is its favourite haunt, although it is found wherever there is low cover by the waterside, and it shifts about the country a good deal in order to find these desirable conditions. It swims and runs on aquatic plants like the little crake, and dives readily if pressed ; but it is shyer, and comes out less into the open, keeping more to swampy places than the open water itself. It is a sociable bird, several being usually found near together, and is also rather noisy, the voice being, according to Hume " a single note, repeated slowly at first, and then several times in rapid succession, winding up with a single and somewhat sharper note in a different tone, as if the bird was glad that the performance was over." This call is chiefly heard during the breeding-season. In feeding this species is less exclusively insectivorous than the little crake, taking wild rice and other seeds freely, as well as greenstuff. The nest is well concealed among rushes, wild rice, or marsh grass, and is made of that sort of vegetation. The eggs number about half a dozen, and have faint but thick dark frecklings on a greenish-drab ground.

Whity-brown Crake.

* *Poliolimnas cinereus* (Brit. Mus. Cat. Birds, vol. xxiii).

This small crake, considerably less in size than a snipe, is recognizable among its kind by its very plain colouring of light brown, shaded with grey in front, above, and white below ; the legs are green. The young have none of the grey shade about the head.

---

* *Porzana cinerea* on plate.

The whity-brown crake is a bird of the Far East, ranging from the Malay Peninsula east even to the islands of the Pacific. In its Malayan haunts it has been observed to prefer cultivated land to the wilds, and is very numerous in autumn in the Singapore paddy-fields, especially in those richly manured with urban refuse. Hume figured it along with the Malayan banded crake owing to its having been supposed to have occurred in Nepal, a mistaken idea based on a wrong identification, for which he was not responsible, and which he discovered after the bird had been drawn.

Corncrake.

Crex pratensis.

The common corncrake or landrail of Europe, which ranges east to Central Asia, and is a great wanderer in spite of its ordinary reluctance to fly when disturbed, is nevertheless very rare in India, its usual winter quarters being in Africa. It has, however, been reported from our area, and was actually once obtained in Gilgit in early October, so that it is worth mentioning that it is rather larger than a quail, with a short bill, and chestnut wings contrasting conspicuously with its streaky-brown upper-parts. It shows the barring on the sides so usual in rails, but the darker bars are only light brown; the rest of the under-parts are plain light brown, the breast and cheeks being grey in the summer dress. This is the only rail really esteemed in England, being very fat and good eating, though several of the family are habitually shot in America and on the Continent. The peculiar double call, well rendered by Bechstein as "*arrp, schnarrp*," is very characteristic of the bird in its summer haunts, but is not likely to be heard in India.

Ruddy Crake.

* *Amaurornis fuscus.*

This little crake, about the size of a quail, resembles the Malayan banded crake in having red legs and to some extent in colour, being chestnut on the face, neck and under-parts, but

---

* *Porzana* on plate.

PORZANA CINEREA ½ _ PORZANA FASCIATA

it has no black-and-white barring, and the back of the head and upper-parts generally are olive-brown : in young birds this brown tint replaces the chestnut on the fore-parts.

Except in the extreme North-west, it is found over India, Burma, and, in winter, Ceylon. It extends through Assam to Pegu and Arrakan. In some parts of India it is, however, little known, being only recorded from Mysore and the Wynaad in the Peninsula.

It is sociable and fond of weed-covered ponds, on the vegetation of which it runs about, coming freely out of the cover on the banks in the early morning, and feeding on the insects to be found on the leaves. It rises readily enough when disturbed, though some specimens prefer to dive and others to run to cover. In more open water it swims about like a moorhen. Besides insects, it eats seeds and greenstuff, and takes grit freely, like rails in general. By day it hides among the fringing herbage of ponds or cultivated land of a wet character.

It breeds among waterside herbage like Baillon's crake, but makes a rather bigger nest; the eggs, to be found from July to September, are about half a dozen in number, with reddish and dull mauve spots on a background of tinted white. Outside India it is found in China and Japan, and in the other direction as far as Java.

Elwes's Crake.

* *Amaurornis bicolor.*

At several thousand feet elevation in Sikkim, as also in the Khasi Hills, there occurs the present species, distinguished from our other small crakes by the large amount of slate in its plumage, which is only diversified by the brown back and wings, all the neck as well as the face and under-parts being grey. The legs are of a pale dull red, and the size is about that of a quail.

The bird has been found in the usual haunts of rails, in cover alongside pools, swamps, and wet rice-fields. It may probably be found to go lower down in winter and to extend some distance along the hills, birds of this kind being generally little noticed.

---

* *Porzana* on plate.

Brown Crake.

**Amaurornis akool.*

The brown crake is a very plain, dingy bird, its dark greenish-brown plumage being only slightly relieved by a white throat, and grey face, neck, breast, and belly; the legs are dull red, or in young birds reddish-brown. The bird in Hume's plate with a black face is a young one which has some of the black down of the nestling coat—almost always black in rails—still remaining on the head, which is not uncommonly the case in otherwise fledged specimens of this species. The brown crake is large for a bird of this group, approaching the grey partridge in size. This is a North Indian bird, ranging along the foot of the Himalayas to the Khasi Hills, and occurs as far south as Mysore; but it is only common in the north. It is not migratory, and breeds from May to September; nesting, it is said, twice during this period. It feeds particularly on small animal life, and is not so much of a skulker as many of its tribe, and early in the morning may be found in the vicinity of water, running about on the bare ground or rocks, and is frequently seen swimming. Pools, streams, and open wells, with but little cover, are frequented by it, and it often perches; it is, indeed, found in the sort of situations a moorhen frequents. It nests in high grass or on bushes, and lays about six brown-spotted pinkish-white eggs. Outside India it is found in China, but has not turned up in the intermediate countries.

White-breasted Water-hen.

Amaurornis phœnicura. *Dawak*, Hindustani.

The contrast between the black upper and white under-parts of this bird, which is besides of a fair size for a rail, being about as big as a partridge though much slimmer, make it a conspicuous object whenever it comes out of its cover. This it pretty frequently does, for it is the most indifferent to human proximity of all our rails, and is quite common, not only in

* *Porzana* on plate.

EURYZONA ♂ CANNINGI

cultivated places, but in the actual neighbourhood of houses and in gardens. It is also, though at home by the waterside and not averse to swimming, not so confined to watery places as most of the family, but frequently seen in hedges and among crops, away from water.

It is not only the most familiar, but about the most widely distributed of all our rails, living nearly everywhere within our limits, even in the Andamans, where it is quite abundant. It is, however, rare in the North-west and does not ascend the Himalayas, though found in the swamps at their bases. Although less timid than rails in general, it has all their essential characteristics—fluttering flight with hanging legs, flicking up of the tail, which in this case displays the chestnut patch underneath, running, swimming, and perching powers, and omnivorous dietary.

The rail habit of being more heard than seen is also very well developed in this species, for it is a very Boanerges among birds, and can literally roar down all the other waterfowl. It generally nests off the ground, on trees, reeds, &c., but makes the usual style of nest constructed by rails, of grass and reeds, sometimes with a twig foundation. It may commence breeding in May, or do so as late as September, according to the district it lives in. The eggs are spotted with reddish-brown and dull, pale purplish on a buff ground, and range from four to twice that number. The down of the chicks is black, and the young birds in their first feather are rusty above and smutty below, while retaining the general pattern of old ones. This familiar bird ranges east to Formosa and Celebes; it has many Indian names: *Boli-kodi* in Telugu, *Tannin* or *Kanung-koli* in Tamil, *Kaul-gowet* in Burmese; while in Oudh it is called *Kinati*, *Kurahi* in Sind, and *Kureyn* by the Gonds.

Moorhen.

Gallinula chloropus. *Jal-murghi*, Hindustani.

The familiar moorhen of home waters is also not uncommon in India, though not so familiar or widely distributed as the white-breasted water-hen; its scarlet forehead-patch and white under-tail feathers, contrasting with its dark plumage, are quite as distinctive of it here as in Europe. Yearling birds, however, are lighter in colour and have no red on the head, though they show the white stern. The Bengali name is *Dakah-paira*, and the Telugu *Jumbu-* or *Boli-kodi.*

Moorhens breed in India, making a large nest among aquatic herbage; the eggs are spotted with chestnut and mauve on a greyish-buff ground, and as many as nine may be laid. The food bill of this bird is as extensive as that of the common hen, insects, worms, herbages, and grain, all being consumed. It is not often eaten, but goes well enough if the rank and greasy skin is removed. The note is very characteristic, a harsh *kur-rek;* the flight heavy and low by day as a rule, though at night the bird travels long distances.

Coot.

Fulica atra. *Dasari*, Hindustani.

The moorhen is nearly as aquatic as the ordinary ducks, and the coot, which is abundant in India and Burma, though absent from Ceylon, bears the same relation to it as the diving ducks do to these, keeping almost constantly afloat and getting much of its food below water; it dives with a spring like the whistling ducks, and especially searches for water snails; it also feeds on weeds, and I have seen one capture a small fish, not by diving, but by suddenly ducking its head under. Grain is also readily devoured if obtainable. Coots are in fact constantly seen in association with ducks in India, and may easily be, and no doubt often are, mistaken for them; but the entirely black plumage, and white bill and forehead-patch are very distinct differences from any duck, and even when these points are not noticeable, the rounded back and small head carried well forward distinguish these swimming rails from the duck family. Coots also rise less readily than most ducks, and though often exceedingly numerous, get up individually and not in flocks. They are often so mixed up with the ducks they associate with that many may be killed by accident; few people would make them a special object of pursuit, as they are not birds to eat when

PORZANA MARUETTA

ducks are obtainable, having a rank, oily skin, and a great tendency to ossification in the drumstick tendons!

Hume, however, says that coots as well as rails and crakes "will furnish a savoury enough dish if, instead of plucking them, you *skin* them and then soak the bodies for a couple of hours in cold water (which should be changed at least twice) before putting them into the stew-pan, with onions, and, if you can get it, sage."

When brought to hand, the coot, if not killed dead, will give plenty of proof that it is not a duck by the vigour of its scratches; the feet are not only provided with particularly strong claws, but are webbed in a curious manner, with a separate scalloped bordering web to each toe, for no rail, not even such a very aquatic species as this, has any web *between* the toes. Coots walk quite well, but are not often to be seen doing so; they nest among the aquatic vegetation, or even on the bottom of shallow water, building the nests up into islands, and using a large quantity of material, chiefly rushes. The eggs are pale buff or drab with copious sprinklings of black, and about as big as hens' eggs; the young chicks are black, but show bright tinting of red, blue, and yellow about the heads. The breeding season is a rather extended one, beginning in May with the birds inhabiting the hills, while in the plains birds are to be found nesting after June; but a large proportion of the coots to be found in India in winter are only migrants from the north, the bird having a wide range all across the old world, and being familiar in Britain among other European countries, though not nearly so common as the moorhen. Its familiarity to natives is attested by its names, *Burra godan* in Burmah, and *Bolikodi* in the Telugu language, while other Hindustani names besides that given above are *Khekari*, *Khuskul*, and *Ari*.

Porphyrio.

Porphyrio poliocephalus. *Kaim*, Hindustani.

The largest of our rails, and distinguished by its blue colour not only from them, but from all our other waterfowl, the porphyrio is a bird which immediately attracts attention by its great beauty, the azure, cobalt, and sea-green plumage being finely set off by the scarlet bill, forehead and legs, and the white under the tail. Some specimens have grey heads, but the presence or absence of this hoary colouring is individual. The bird is often called purple moorhen or coot, but differs from both in several points besides size and colour, notably in the great thickness of the beak, with which it can give a very severe bite, and in the curious habit, so remarkable in a waterfowl, of using its foot to hold its food, chiefly vegetable—like a parrot. Although always found near water, it does not swim much, and has the true rail love for cover; it perches freely and climbs well among the reeds.

It is found all over our Empire, and of late years has even been recorded as far west as the Caspian; it is resident, and nests during the rains; the eggs are about the same size as the coot's, but richer in colour, having a reddish tinge both in ground-colour and spots. The other Hindustani names of this favourite bird, which is not usually shot, though many specimens are sent to Europe alive, are all apparently variants of that given above—*Khima*, *Kharim*, and *Kalim;* in Ceylon the names are *Indula*, *Kukula*, *Sannary*, and *Kittala*.

Water-cock.

Gallicrex cinereus. *Kora*, Hindustani.

The water-cock, as it is to be met with in the shooting-season, is a game-looking bird with light brown plumage, diversified by streaks on the back, and bars on the under-parts of a darker shade. It has the usual long legs and toes of a rail, and a leaf-shaped bare patch on the forehead. Although much lighter in build, the male is nearly as big as a coot, the hen being little larger than a moorhen—a sex difference and unique among the rails, as is also the male's assumption of a striking nuptial dress; in this attire he is of dull black on the head, neck, and under-parts, while the bare patch on his forehead, which, like the legs, is red, swells up until it becomes at the end a pointed horn. The female has legs of a dusky green.

½

PORZANA PARVA

The kora, as this bird is generally called, is widely distributed with us, but although it ranges as far north as Japan outside our area, it keeps in our Empire to the warmer districts; it is a thorough marsh-bird, but seems to be rare in some localities where it was formerly common, for Bengal was credited with harbouring plenty of the species, and yet I never saw half a dozen specimens during the whole time I was in Calcutta. The kora is quite a good table-bird, so that if it is getting scarce this is a pity; but being nocturnal, it is not likely to come under notice in the same way that the diurnal coot and moorhen do.

The breeding-season is during the rains, and the eggs are greyish-buff with mauve and chocolate spots; the nest is among aquatic herbage. Besides *Kora, Kengra* is Hindustani name for this bird; in Ceylon it is called *Willikukulu, Kettala,* or *Tannir-koli,* while the Burmese name is *Boun-dote.* Its familiarity to natives is no doubt due to the fact that in some districts of the North-east it is reared by hand and kept as a fighting bird.

Sarus Crane.

Grus antigone. *Sarus,* Hindustani.

Hume quite rightly says that this bird is not properly a game bird at all, but simply comes in as a relative of the cranes which may be so reckoned; and this is just as well, for it is one of the most conspicuous and ornamental birds in the country. A "common object of the wayside" to the traveller by rail, its tall grey figure, about five feet high, surmounted by the bare scarlet head, cannot escape observation. Almost invariably a pair are seen together, and the hen can be distinguished by being about a head shorter than her mate, who is about five feet long, and stands about as high; for being a bird of very erect carriage the sarus looks all its size, and appears to be the biggest bird in India, though the great bald adjutant stork (*Leptoptilus dubius*) exceeds it in measurements, and the great bustard, no doubt, in weight.

It is worth noting that the neck in this species, just below the bare scarlet part, becomes white in the breeding season, and the long wing-plumes also get whiter then; for the existence of this white in the plumage, and the general paler tone of the same, are the chief distinctions between the Indian and Burmese types of this crane. The sarus (often miscalled cyrus!) is practically purely an Indian bird, and is not known to occur in Transcaspia and Persia, though, curiously enough, sometimes turning up in Russia. Even in India it is far from being universally distributed, for it does not range into the hills, except in Nepal, where, according to Hume, it has been introduced. Nor does it occur in Mysore or any district south of this, while it is rare in Sind. In the open country of northern India it is well known in all well-watered districts, and rather prefers cultivated land; it is extraordinarily tame for such a large bird, but this is due to the fact that it is very rarely molested; its flesh is not esteemed, although the liver is good, and natives do not like its being shot, as they admire it, although not considering it at all sacred.

In case there is any real reason to kill so harmless and beautiful a bird, the pair should both fall together, for there is told about this bird the same tale that is related of the little parrots known as "love-birds," that if one is killed the survivor dies of grief. Love-birds do not always do this, nor does the sarus; generally, as Hume says, after haunting the scene of its bereavement for some days or even weeks, and calling continually, it disappears, "and," he says, "it is to be hoped, finds a new mate, but on two occasions I have actually known the widowed bird to pine away and die: in the one case my dogs caught the bird in a field where it had retreated to die, literally starved to death; in the other the bird disappeared, and a few days later we found the feathers in a field where it had obviously fallen a prey to jackals." No doubt, many birds having pined till they cannot recover, fall victims in this way; a healthy sarus has little to fear from vermin, at any rate if there is water in which it can more readily stand on its defence. Dogs are easily beaten off from the great nest, which is a sort of artificial island in many cases, built up on a rise in the bottom of some bit of water, where half a foot to two feet of foundation may have to be laid before the nest rises above water, though, of course, actual islets

CREX BAILLONII.

Hatton Garden, London.

are also selected. The nest is made of reeds, rushes and straw, and is raised more or less above the water according to circumstances, the egg-bed being about a foot out of it. In times of rains the birds raise the nest; in fact, their nesting proceedings are much like those of the familiar tame swan at home. Sometimes the nest is built among high reeds, on a platform of these bent and trodden down.

They seldom show fight when their home is invaded, but Hume records a case in which a hen brooding eggs, one of which was actually hatching, stayed on the nest making ferocious digs at a native sent by him to investigate, till he had to flap her in the face with his waist-cloth to get her off; and Mr. D. Dewar, in his book, "Glimpses of Indian Birds," describes how, when a man of his captured a chick, the cock bird deliberately stalked them, and approached within four feet, only to be driven off by hostile demonstrations. His description of the chick is worth quoting: "It was," he says, "about the size of a small bazaar fowl, and had perhaps been hatched three days. It was covered with soft down; the down on the upper parts was of a rich reddish-fawn colour, the back of the neck, a band along the backbone, and a strip on each wing being the places where the colour was most intense; these were almost chestnut in hue. The lower parts were of a cream colour, into which the reddish fawn merged gradually at the sides of the body. The eyes were large and black. The bill was of pink hue and broad at the base where the yellow lining of the mouth showed. The pink of the bill was most pronounced at the base, fading almost to white at the tip. The legs and feet were pale pink, the toes being slightly webbed."

Even when the young bird is fledged the head remains covered with this chestnut down for a time; the beak in adults is dull green, as is the scalp, but the legs are always pink, though the eyes become red. The wings do not fledge till the bird is of a good size, and the old ones, at any rate in captivity, lose all their quills at once, like geese, when moulting, so that they must depend on fighting enemies rather than flight during this season; but no doubt they seek localities where defence is easy.

At the best of times they fly but little; if there be nothing such as a fence or copse to hide a possible enemy, they will rather walk a mile or two than fly, and when on the wing do not rise above twenty yards even in a five-mile flight, according to Hume. No doubt, however, their powers of flight are capable of far greater exercise, or they could not get so far as Russia. The call of the sarus is very characteristic, and the male and female sing, as it were, together. First the male, raising his head and bill perpendicularly, and lifting the wings at the elbows without spreading them—much like an angry swan—gives out a loud single note; the hen instantly follows, the cock replies, till the appalling duet, which can be heard two miles, is finished. It will be gathered from what has been said that the sarus is a pairing rather than a flocking bird, but the young remain sometimes with their parents; as two or even three eggs are laid, they should make up a little flock, but, as a matter of fact, often only one young bird is reared, a result to which the numerous birds of prey probably contribute, in spite of the watchfulness of the parents, both of which carefully attend the young ones; these are active, not helpless nestlings.

The eggs are very large, long, and hard-shelled; they vary, but may be nearly four and a half inches in length. They are spotted with pale yellowish-brown and purple on a white, pale sea-green, or cream-coloured ground.

The food of this crane is sought either on land or in shallow water, but it is less of a marsh feeder than our other species, and spends more time out of water than in it as a rule, except when nesting. Small animals, such as lizards, frogs and insects, form a large proportion of the food, though much is also vegetable; and in captivity the bird readily eats raw meat as well as grain.

The only native name that needs be noted in addition to the ordinary one—*Sarus*—is the *Khorsang* of the Assamese, in whose country the bird finds its eastern limit.

½

PORZANA _ BICOLOR & FUSCA.

Burmese Sarus.

Grus sharpii. *Gyo-gya*, Burmese.

Although nowadays classed as a distinct species, the sarus of Burma differs very little from the Indian bird, being merely darker grey, with no white anywhere; it has a dingier aspect altogether, and is inclined to be smaller, while the hairs about the throat are very scanty.

This is the large crane, not only of Burma, but of the Malay Peninsula, Siam, and Cochin-China, and the older accounts, such as those of Hume, of the sarus occurring in these countries, must be taken as referring to this species; but the common sarus is the crane of Assam, judging from a skin in the Indian Museum in my time, which I was able to compare with another of the present form from Upper Burma, also in the collection.

Mynheer F. Blaauw, in his valuable monograph on the cranes, gives an interesting account of the breeding of this bird. He says: "The Eastern sarus crane has been found breeding in the months of August and September, and it probably also nests later in the year, as Davison found young birds in Burma, still unable to fly, as late as December. Wardlaw Ramsay, who records its breeding near Tonghoo, tells us that, although he did not find the eggs himself, eggs were brought to him by the Burmese. They described the nest as a pile of weeds and mud, situated generally in the midst of a swamp. On September 29, a Burmese brought him an egg and a newly hatched chick . . . the little bird was given into the charge of a common hen with doubts as to the result. She, however, took the greatest care of it, and showed great wrath if anyone attempted to touch it. On the morning of the eleventh day, however, the little creature died. When just out of the shell it devoured worms greedily."

Davison found that the young birds displayed great cunning in taking cover, but would resort to the plan traditionally ascribed to the ostrich, of hiding their heads when fairly run down in the open. These birds were destructive to the young plants in paddy nurseries, and he never saw them eating anything else. They themselves were considered a great luxury by Davison's friends in Moulmein, to whom he used to send them. It may be that it is on account of being shot for food, although the Burmese do not like them being killed, that the disposition of this race of sarus is different from that of the Western form; it is shy and wary, needing to be approached by a bullock cart, or in the rains by a canoe. The hen has a silly habit of standing on top of her nest at daylight, and calling—a proceeding calculated to give away her family affairs. The eggs of this sarus appear to run lighter than that of the other, having only a few rufous blotches, or even being all white.

But the only thing really distinctive about the habits of this bird is that it is to some extent migratory, assembling in numerous bands and taking long and high flights. Anderson, at Ponsee, saw them passing in V-shaped flocks in the direction of the Burmese valley, flying so high as to only appear as specks. Nine such flocks, each numbering about sixty birds, assembled above the high mountain where he was camped, and commingled, with aerial evolutions, breaking up into two masses, and then into the V-formation again in smaller groups. Nothing like this is ever seen with the Indian sarus. Davison also saw bands, numbering up to sixty birds in each, arrive near Thatone in August; there is evidently a good deal to be made out about the migration of this bird, as in the case of so many tropical species wrongly believed to be stationary.

Common Crane.

*Grus communis.** *Kullung*, Hindustani.

One of the points in which India recalls classical times in Europe is the yearly winter visitation of the common crane, an enemy to the farmer, just as it was in the time when Æsop's fables were written. Everyone knows the fate of the misguided stork whose virtue did not save him when caught with the cranes, and Virgil complains of cranes as well as geese in enumerating the troubles of the Roman agriculturist.

At home the crane is now the rarest of visitants, and the common heron often usurps its name; and as this bird is found in

---

* *cinerea* on plate.

½
PORZANA AKOOL

India too, it may be pointed out, for the benefit of beginners, that though both are big tall grey birds, the crane may be distinguished on the ground by the long curved plumes which look like a tail, but really grow on the wings, and especially on the wing by the neck being extended, as well as the legs, herons always drawing the neck back when they fly.

When near at hand—which a crane is not likely to be, if healthy—it will be seen that it is a much bigger bird than the grey heron, nearly four feet long in fact, and has no crest or breast-plumes, but a bald red patch on the head. The sober grey of the whole of the body-plumage is only relieved by more or less black on the ends of the wings, and by bands of white along the sides of head and neck. The sexes are alike, but the young of the year can be distinguished by a mixture of buff in their plumage, especially on the head and neck, and their less developed wing-plumes.

The bird in the plate, by the way, is much too dark and dull a grey, and has been given a well-developed hind-toe like a heron's, whereas this toe is really very small and quite useless, cranes, at any rate our Indian species, not being perchers like herons. They are also much more sociable, being always in flocks, usually ranging in number from a score in the south, where the birds are nearing the limit of their range, to several hundred in the Northern Provinces. This crane's southern limit appears to be Travancore, and its special haunts are the Northern Provinces of our Indian Empire, while it is not known in Burma or Ceylon.

These cranes may come in as early as August, in Sind, but as a rule October is about the time of their arrival; most go away in March, but some may be found even in May at times. They haunt open places and the vicinity of water, preferring rivers to tanks, but feed much away from the water, as a large part of their food while in India consists of various crops, especially wheat, grain, pulse and rice, for cranes are mixed feeders, not purely animal feeders like storks and herons. Early morning is their chief time for raiding the fields, and they do a great deal of damage, devouring not only the grains and pods of the cultivated plants, but the young shoots. They will also attack sweet potatoes, water-melons, and other vegetables. Dál is about their favourite of all crops, and where this grows higher than they are, they are more easily got at than is usually the case, since they cannot see the foe approaching in the distance.

In the ordinary way they are as wary as most large birds, and take careful stalking, always having sentries on duty when feeding; they are, however, particularly well worth pursuit, as not only are they nuisances to the farmer, but excellent game when obtained, always provided they have had time to eat enough of the vegetable food most of them prefer to get rid of the coarse flavour resulting from the diet of animal small fry they have been eating before the crops are available. This crane is, in fact, one of the delicacies of the classical and mediæval cuisine which is really worth eating; this being more than can be said for a good many of the fowl our forefathers used to relish so much—in days, be it remembered, when fresh meat during at least half the year was very hard to come by.

At night cranes resort, if possible, to an island sandbank to roost, where they sleep standing on one leg. This is, no doubt, as a protection against four-footed enemies, although such vigilant birds are not very likely to be surprised by such foes. Few birds also will attack this powerful species, and Prince Mirza, in his valuable and interesting book on hawking, translated by Colonel Phillott, says that if you want a falcon to take cranes, you must not let her fly at herons, these being so much easier game. He also says that if one member of a flock is brought down by the hawk, its companions will all come to its assistance, and much commends their *esprit de corps.* Wounded cranes, by the way, run fast and swim fairly well, while they are nasty customers to tackle without a stick.

Their trumpeting note is very fine and characteristic, and, in addition to their habit of forming lines and wedges in flight, has always made them conspicuous; as Dante says:—

> "And as the cranes go trumpeting their call,
> Trailing their long-drawn line across the sky."

And one of the classical crane stories is of the poet Ibycus, who, done to death by highwaymen, called with his dying breath on a passing flight of cranes to avenge him. The story

5

GRUS CINEREA

says the birds did not forget, but some time after were seen circling and calling over a market-place in which the robbers were at the time. One conscience-struck ruffian cried out to his friend, " There are the avengers of Ibycus," and thus betraying his secret, brought justice on the whole gang.

Ranging practically all over Europe, though chiefly breeding in the north—including England once—this great bird has naturally left a very marked impression in literature; it breeds all across northern Asia also, and winters in China as well as India. No nest has ever been found in our limits ; the eggs and young are much like those of the sarus, but smaller.

The native name *Kullung* is generally used also by Europeans; a slight variant is the Deccani *Kullam*, and *Kooroonch* is another Hindustani name ; in Manipur the name is *Wainu*.

Hooded Crane.

Grus monachus. *Nabezuru*, Japanese.

This very rare visitant is distinguished from all our other cranes by the complete and conspicuous whiteness of its head and neck, contrasting strikingly with the body, this being of a darker grey than that seen in any of our other cranes. In form and in having a bald red patch on the head, it resembles the common crane or coolung, but is a little smaller in size, not exceeding a yard in length. Young birds have the grey of a brownish cast, owing to the feathers being edged with brown.

The only record of the occurrence of this bird, which ranges, according to season, from eastern Mongolia and Siberia to Corea and China—sometimes also to Japan—is one by Mr. E. C. Stuart Baker, in one of his articles on the " Birds of North Cachar," published in volume xii of the *Journal of the Bombay Natural History Society*, under the name of " King " crane (*Grus monarchus*). No such species exists, but he evidently meant the present bird. He says : " In December of 1889, whilst fishing in the Mahar River, seven huge cranes flapped overhead down the stream and settled in a shallow pool some four hundred yards away. They at once struck me as being something I had not seen before, and I followed them up, and though I failed to bring my bird down with the first barrel I knocked one over as they rose with the second. He half fluttered and half ran down the stream, and it took a third barrel to bring him to bag; but when it was at last brought to hand, I found myself in possession of an undoubted *Grus monarchus*. The anterior crown was black, otherwise the whole head and neck were white. The brown margins to the feathers of the upper-part made the plumage appear to be a brown-grey. The wing measured full twenty inches."

This measurement would be taken from the pinion-joint to the tip, and does not indicate a "huge" bird, but is correct for this species. Cranes are rare in Cachar, and of other species Mr. Baker only records the sarus, and that only as a pair of casual visitants, so no doubt any crane would reasonably have appealed to him as a huge bird. These details are worth giving, because the specimen was unfortunately not kept. "I was three days from headquarters," says Mr. Baker, "but I thought special messengers would get it in in time to skin, but alas! when I arrived three days later I found it had not been brought in, and the messenger, when questioned, said, 'Oh, it began to smell, so I threw it away.'" It is a pity the attempt to send it on was made, as the head and neck, however roughly preserved, would have been sufficient for identification.

Hume also mentions, in volume xi of "Stray Feathers," what was probably an occurrence of this species in Manipur. "On March 13, when between Booree Bazaar and Bishnoopoor, a small flock of cranes passed me at a distance of about two hundred and fifty yards, flying low and due north. I got on to a small mound and watched them for probably more than a mile with my glasses, but when I lost sight of them they were still flying steadily away northwards. Now, whatever they were, they were certainly none of our Indian species. . . . They were of a uniform dark hue, much darker than *communis*, and had the whole head and upper-parts of the neck pure white. Of course, one says at once '*Grus monachus* no doubt.' But so far as I have been able to study the distribution of this group it is simply *impossible* for *monachus* to be in Manipur in March. I never

1/5

GRUS LEUCOGERANUS

saw the birds on any other occasion, and I do not pretend to know what they were, beyond this, that they were cranes of the *monachus* type and probably some undescribed species." No such species has ever turned up, and of course the argument as to date and locality has no value in the case of strong-winged migrants; there can be practically no doubt that Hume's birds were simply hooded cranes. Not much is known about the bird anywhere; its eggs have not been taken yet. It travels in small flocks, and arrives at its breeding-grounds in the north in April as a rule, and leaves for the south in August. Although it is rare in captivity, the London Zoo has a fine pair at the time of writing. I can see no brown on their plumage, and I notice that they wade a great deal.

White or Snow-wreath Crane.

Grus leucogeranus. *Karekhur*, Hindustani.

In height and length being only by a few inches less than the sarus, this splendid snow-white bird can easily be distinguished from anything else in India if seen where the size can be appreciated, and if this is not the case, still its pinky-red face and legs will distinguish it from a large egret or a spoonbill. From the white stork (*Ciconia alba*), also red-legged, the apparent absence of black in the plumage will distinguish it, while though when on the wing the black pinion-quills are conspicuous, they should not lead to confusion with the stork, which has nearly all the wing as well as the tail black.

Young birds of the year are still more unmistakeable, being buff in colour, at any rate when they first arrive. Such birds are generally found along with the two parents, for the white crane, like the sarus, is essentially a lover of family life; the flocks of half a dozen or so sometimes seen appear to be young two-year-old bachelors and spinsters, and no doubt such, with a sprinkling of bereaved old birds, make up the larger flocks which now and then occur.

This crane is purely a winter visitor, and a rather local and scarce one at that; though, judging from the numbers the dealers used to get hold of when I was in India, at any rate for several years following 1894, it is liable to come in some years in considerable numbers. The districts affected by it during its stay, which is between October and March, are all in the North-west, from Sind to Oudh, in which latter province it is called *Tunhi*. It is very local and very aquatic, being almost always seen in the shallow water of jheels and marshes, where Hume found it fed exclusively on vegetable food, the bulbs, seeds and leaves of various water-plants, especially rushes. The parents displayed the greatest affection for their young, pluming its feathers, and calling it to eat whenever they found a promising rush-tuft, while if it were shot they would circle round in the air for hours, calling disconsolately, and would return to the spot for days afterwards.

The call of this crane is much weaker than that of our other species, "what," says Hume, "for so large a bird, may be called a mere chirrup." But, like the sarus, it has a sort of set song, to which the term chirruping can hardly be fairly applied; the attitude in which this note, which is like a more refined and musical edition of that of the sarus, is given forth is peculiar. At first the bird begins to call with the bill bent in towards the breast; with each note the bill is jerked further forward, while the wings are lifted and the pinion-quills drooped exposing their blackness, till, by the end of the song, the bird is calling with bill and neck erect, in the typical sarus position.

This is a very wary bird, and when obtained is not good eating, while it is not a devourer of crops, so that there is no particular reason to trouble about shooting it. Its breeding-home is in Central Asia, Siberia, and Mongolia, and there its feeding habits are probably different from its vegetarian practices in India, for in captivity in England it readily eats fish, and will wait to catch them like a heron, and devour young ducks; it also digs for earth-worms.

Eggs taken in the wild state are still a desideratum, but several pairs have laid and sat in captivity, though up to date no young have ever been hatched. One pair in the London Zoo nests in this futile way year after year; the eggs are two in number, and olive-brown in colour with dark brown blotches.

ANTHROPOIDES VIRGO.

Mr. R. Cosgrave, in some interesting notes on the cranes at Lilford Hall in the *Avicultural Magazine*, says that the white cranes kept there are miserable in heat and rejoice in cold; and, though this is not the case with the Zoo birds, which always behave and look much the same, it is quite possible that, as he suggests, the climate accounts for the infertility of the eggs so far produced in England.

Demoiselle Crane.

Anthropoides virgo. *Karkarra*, Hindustani.

The demoiselle crane is the smallest species found, not only in India, but anywhere; it is not quite a yard long, and so would be more likely to be mistaken for the grey heron than is the common crane, were it not that the adults have their grey plumage strikingly set off by the black face, neck and breast, and long white plumes drooping from the cheeks; while in the case of the young, which have only black on the neck, and but a little there, and the "kiss-curls" only just indicated, the shorter beak and neck outstretched in flight are sufficient distinctions.

Moreover, demoiselle cranes are, even more than the common crane, likely to be found in flocks; they are extremely sociable, and some of their assemblages are enormous. Captain E. A. Butler says: "I have seen tanks fringed with a blue margin of these birds at least sixty yards wide, and extending over several acres of ground, over and over again." This was in Guzerat, and here, as well as in Kathiawar and the Deccan, are the bird's headquarters during its stay with us, for it is only a winter visitor, generally leaving in March, but sometimes waiting till May; the month for arrival is October. Besides the provinces named, the demoiselle also visits North-western India generally, and penetrates as far as Mysore in the Peninsula; but in Lower Bengal and the countries to the eastward it is not found, though occurring in China in the winter, and in the end of the Peninsula is a rarity, while it does not reach Ceylon. It is called *Kullum* in the Deccan, but wrongly, as this name seems to apply properly to the common crane or coolung, unless it simply means "crane"; the Mahratta name *Karkuchi,* the Canarese *Karkoncha,* and the Uriya *Garara,* are evidently, like "karkarra," an attempt at imitating the note, which in this species is very harsh and grating, quite at variance with the dainty grace of the bird, which well merits the name of "demoiselle."

It is a cheerful, playful bird, and in some districts spends most of the day on the wing, soaring round and round in circles, apparently merely for exercise. At such times it is most difficult to get near, and is, generally speaking, a very wary and thoroughly sporting bird. It is also excellent eating, at any rate when it has had the chance of feeding on cultivated produce, to which it is as partial as the common crane; for this species also is, in its winter quarters at least, by preference a vegetable feeder. A favourite food is the *karda* or safflower seed, but it eats grain freely, and thrives well on it in captivity. Young reared in Europe in captivity, however, were fed by their parents on insects at first.

After feeding on land they betake themselves to the edges of large tanks, and especially rivers, and roost in large flocks in such places, or in open plains, with sentinels set, the roosting flock breaking up into detachments with daybreak, when they fly abroad for food.

The breeding range of the demoiselle is very wide, from Southern Europe eastwards all through Asia, but in temperate regions always, for this species is at all times a less northern bird than the common crane. The nest is on the ground, but made, curiously enough, of pebbles, with which also all the inequalities of the ground round about are filled in. The eggs, two in number, are much like those of the common crane, but smaller, and with more distinct markings on a darker ground.

It is worth mentioning that in Southern India some sort of sanctity attaches to this bird; patches of crops are left for it to feed on, and in Brahmin districts one may have serious trouble for shooting one, unless feeling about such matters has altered since Hume wrote a generation ago.

EUPODOTIS ♀ EDWARDSII

The Great Indian Bustard.

Eupodotis edwardsi. *Hukna*, Hindustani.

The largest and most esteemed of Indian game-birds, this fine bustard is easily recognizable; in size it exceeds the ordinary domestic turkey of India, and its long neck and legs make it conspicuous. Its colour is dull brown above, white below; in the old male the neck is also white, but in hens and young cocks this part appears grey, owing to being pencilled over with black. In any case the crown of the head is black, contrasting strongly with the light cheeks.

On the wing this bustard looks not unlike a vulture, moving with slow heavy sweeps of the wings, but it flies low and never sails.

Old cocks—and these alone ought to be shot—are distinguishable not only by their white necks, but by their much greater size, as they are twice as big as hens, often weighing twenty pounds, and it has been said sometimes even twice as much. In length the cock is four feet, the hen about a foot less, while the wing expanse is about double the length. The great Indian bustard is a purely Indian bird, and is still found in the same districts as it frequented in the days of the pioneers of Indian ornithology.

In Ceylon and the extreme south of India it is not found, nor in the eastern portions, Bengal, Behar, Chota Nagpore, or Orissa. Needless to say, it does not extend into Burma or the Malay countries; but, curiously enough, the Australian bustard (*Eupodotis australis*), commonly known in the Commonwealth as "plains turkey," or simply as "wild turkey," is so very similar to the Indian bird that it can hardly be regarded as anything but a local race of the same species, although one would have expected to find a bustard of any sort in Australia about as much as a cockatoo in India.

Presumably the bird once ranged throughout the intervening countries, but these became unsuitable for it owing to a change in conditions; probably the growth of forest, for even in its chosen haunts in the plains of the Peninsula of India this bustard, like most of its family, is essentially a bird of the open, and avoids heavy cover.

Dry undulating land, bare or grassy, is the bustard's favourite country, but when the grass in its haunts is cleared off it will resort to the waterside, or depart altogether to a locality where there is more grass. It also frequents wheatfields, and will eat grain as well as other seeds, shoots and berries, especially those of the ber and caronda, though it is by nature rather an animal than a vegetable feeder, especially relishing grasshoppers. Beetles—including blister-beetles—caterpillars, and even lizards and snakes, form part of its food; no doubt it will in practice eat any small living creature it comes across, as the great bustard of Europe does.

Although this bustard does not fly high, it rises easily and is willing to travel several miles at a time, and it must traverse considerable distances at times in changing its quarters in search of suitable feeding-grounds. No one in modern times has ridden it down, as a writer in the old *Bengal Sporting Magazine* said he had known done; perhaps a bird in heavy moult might succumb to persistent hunting, but the pace of this large species on the wing is much greater than it appears, and would not give a horseman much chance to come up with it and tire it out.

Generally speaking, it is considered a most difficult bird to bring to bag, requiring very careful stalking, though now and then birds surprised in cover taller than themselves may fall easy victims. Now and then a few old cocks will associate with blackbuck, no doubt for the sake of mutual protection by watchfulness, just as the true ostrich in Africa associates with the zebra and gnu, and the rhea, the so-called "ostrich" of southern South America, with the guanaco or wild llama. At all times this bustard is commonly in some sort of company of its own kind. A few old cocks or hens may chum together apart from the other sex when not breeding, while in the breeding season a strong cock collects about him as many as half a dozen wives.

Nowadays, however, flocks of as many as two dozen birds, such as Jerdon records, are hardly ever to be seen, the largest parties generally numbering under a dozen.

In courting, the male of the Indian great bustard goes through an extraordinary display. Strutting about with head lifted as high as possible, he cocks his tail, inhales air in repeated puffs, expanding and contracting his throat, and at last blows the neck out into a huge bag till it nearly reaches the ground, when he struts about displaying this goitre, and with his tail turned over his back, at the same time snapping his bill and uttering a peculiar deep moan, no doubt the origin of his Mahratta name of *Hum*. His ordinary alarm call is a most unbird-like noise which strikes some people as like barking, while others compare it to a bellow or the distant shout of a man. Hence is derived the Hindustani name *Hookna*, while the Canarese *Ari-kujina-hukki* means "the bird that calls like a man." Captain C. Brownlow also records (*vide* Mr. Baker) "a sort of cackle" uttered by an undisturbed flock.

The breeding season of this bird is extended over more than half the year, Mr. Baker recording eggs taken in every month except December, February, and March, but the main breeding months appear to be from August to November, while the time is locally variable. Only one egg is laid—at any rate as a rule—and this in a slight hollow in the ground, with no attempt at a nest except sometimes a few bits of grass. It may be in the open or in high grass, preferably the latter.

The egg is thick-shelled, and, though variable, tends to a long shape, often over three inches long. It is spotted, more or less distinctly, with brown on a ground of pale brown, dull olive green, or even grey. The down of the chick is buff above and white below, variegated with black on the buff portions.

As this bird increases so slowly, it certainly needs watchful protection, but it seems not to have been seriously reduced in numbers during the last half-century. Although Sterndale records a tame specimen as killed by his pet mongoose, the size of the bird must protect it against small vermin as a rule, while it is on occasion a plucky bird, a correspondent of Mr. Baker's having been actually charged by a winged cock, and obliged to give it another shot. Moreover, the extreme wariness above alluded to is a great safeguard. Some credit the bustard with a keen sense of smell, which may partly account for the difficulty of approaching it.

The flesh of the great Indian bustard is coarse in the case of old cocks, though young birds and hens are better. Such a striking bird has naturally many names. In addition to those given in the text, may be noted those of *Tokdar*, the usual title given by Mohammedan falconers, which is a variant of the *Tugdar* of the Punjabis; in the Deccan we come across the names *Mardhonk*, *Karadhonk* and *Karbink*, while *Sohun* and *Gughun-bher* in Hindustani are used as well as *Hukna* and *Yere-laddu* in Canarese; the *Bat-meka* of Telugu and *Batta mekha* of the Yanadi are evidently allied titles, while a quite different one is the *Kanal-myle* of Tamil.

The Florican.

Sypheotis bengalensis. *Charas*, Hindustani.

The florican, so celebrated for the delicacy of its flesh, is about the size of a peahen, but longer and leggier in build and shorter in the tail. The hen is buff mottled with black, producing a general brown effect; but the cock is very conspicuously coloured, mostly glossy black, but with the wings white, making a most conspicuous contrast in flight, and noticeable even in repose.

The back shows the partridge brown of the hen, and young cocks have hen plumage in their first year, but after the second year are fully coloured; in the intermediate plumage the white on the wing is present, so that a brown bird, if with white wings, may be safely shot as a cock; hens should always be "let off."

Hens are bigger than cocks in this species, though the difference is more in weight than in measurements, a hen weighing four or even five pounds, while a cock will be a pound less as a rule.

This florican is a characteristic bird of eastern Bengal, whence the name "Bengal Florican" often given to it; but in addition to those parts of Bengal which lie north of the Ganges, it is found in the adjoining parts of Oudh and the North-west Provinces: it is well known in the Assam Valley, but not found in southern or western India, or in any country outside India proper.

SYPHEOTIDES ¼ BENGALENSIS.

It is a bird of the open grass country, where it lives solitary, preferring thin grass, though it will take to high thick growth if there is no other cover available; in thick cover it lies close, but in thin short grass it is hard to get near and runs fast and far.

When flushed it flies slowly, but with frequent wing-beats, and generally for only a mile or less, and succumbs to a comparatively slight blow. Like so many other solitary birds, it is noticed to affect particular spots, these being soon after reoccupied when the specimen found haunting them has been killed.

The greater part of the florican's food is vegetable, including sprouts, seeds, and runners of grasses, berries, mustard-tops, milky-juiced leaves, &c., but it takes a great deal of animal food also, feeding particularly on locusts when these can be had, besides grasshoppers and beetles. Corn it does not seem to care for. In the season when blister-beetles abound it feeds freely on them, and is then a very undesirable article of food, as these insects have the properties of cantharides, and a corresponding effect on those who partake of the bird which has eaten them. In the ordinary way, however, the florican is prized as the finest of Indian game birds for the table; its flesh is of high flavour, with a layer of brown without and white within.

The breeding customs of this bird are peculiar; the sexes do not live together, but in the time of courtship, that is to say from March to June, the cock makes himself conspicuous by rising perpendicularly into the air some ten or fifteen yards, with flapping wings and a peculiar humming note; sinking down, he rises again, and so for five or six times, until a female approaches him, for at this time the sexes, though not actually associating, tend to draw near together. He then displays on the ground, with erected and expanded tail, still repeating the humming sound. His affections are very transient, for he takes no more notice of his temporary mate.

For her part, she seeks thick grass cover, and lays two eggs at the root of a grass clump, with no nest. The eggs are about the size of small hen's eggs, of a more or less bright olive-green spotted with brown; one is generally larger and richer coloured than the other. The hen sits for a month, if she is not disturbed, for, according to Hodgson, she is so suspicious that if the eggs are found and handled she is sure to discover it, and then herself to destroy them.

The young are runners, like those of other bustards, and can fly in a month; they stay near the mother, however, till, when they are nearly a year old, she drives them off. As two hens often breed together, and apparently pool their broods for mutual protection, just as eider-ducks do, coveys, so to speak, of half a dozen birds may be found, in contra-distinction to the usual unsociable habits of this species.

With the exception of the humming courting-song of the florican, its only other note is the alarm call of "chik-chik," shrill and metallic, but also uttered in a softer form when the bird is at ease.

The bird is often called "houbara" by sportsmen—quite mistakenly of course—for the houbara, though also a bustard, is quite a different bird, haunting the desert tracts just where the florican is not found. Variants of the Hindustani name are *Charat* and *Charj*, and in many parts of the Terai the sexes are distinguished by name, the male being *Ablak* (pied), and the hen *Bor*. In Assam the bird is the "grass peafowl," *Ulu Mor* of the natives.

Lesser Florican.

Sypheotides auritus. *Likh*, Hindustani.

Better known perhaps by its Hindustani name, this beautiful little bustard is very distinct from all other birds. So long in neck and leg is it that it looks like a miniature ostrich when on foot; its size is only a little larger than that of the common partridge. On the wing it resembles a duck somewhat, having a rapid flight and similar-sized wings to a duck's.

The hen is of the same partridge-colour, a buff mottled with brown, as the hen large florican; and the cock is, like that of the large species, a black-and-white bird, with a partridge-brown back, but his head and neck are closely feathered like the hen's,

Eupodotis edwardsi.

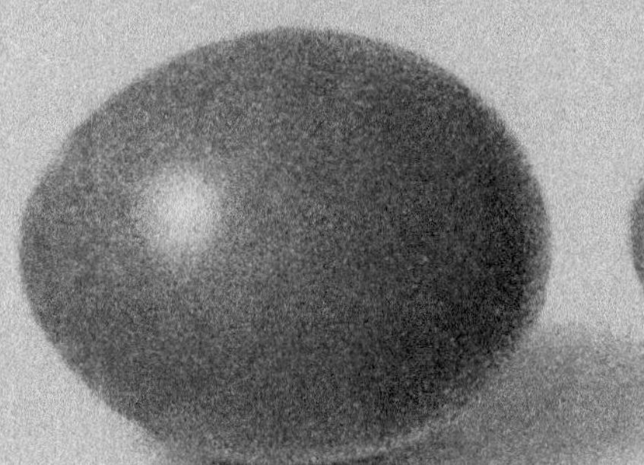

Syphectides aurita

Syphectides aurita

Eupodotis edwardsi

whereas the feathering on this part in the big florican is full and bushy. But the likh has his own decoration in the shape of three long and very narrow feathers, mere shafts, with a tassel of webbing at the tip, on each side of the head. Nothing like this is found in any other bird except some of the black birds of paradise of the genus *Parotia* from New Guinea.

The likh male has not so much white on the wing as the large florican, and he goes into hen plumage for the winter, still, however, retaining a white wing patch. In this species also the cocks are smaller than the hens, and to a greater extent; the cocks weigh about a pound, the hens half as much again.

This interesting bird is one of the many fascinating species which are purely Indian; it is not even found in Ceylon, and though its range in India is wider than that of the large florican, it does not cover the whole country, its real home, according to Hume, being the drier portions of the Peninsula. As, however, it is irregularly migratory, and, like migratory birds generally, turns up individually as a straggler, it may be found almost anywhere, at times in open plain country. There is, however, a general movement north and west during the rains, when the birds breed, and after this they drop back southwards; but the passage is so irregular and dependent on climatic conditions, that the birds cannot be looked for with certainty year after year in the same localities.

This has rather encouraged the iniquitous practice of shooting them during the breeding season, a poaching trick rendered unfortunately easy by the peculiar display of the cock, which at this time springs about a couple of yards from the grass with a frog-like croak, sinking again parachute-fashion with outspread wings. This is repeated about every quarter of an hour, and no doubt attracts the hens, to say nothing of rivals, for cocks have been seen fighting desperately. The hen also springs at times, and the cock may do so without calling. The likh affects cover more than any other of the Indian bustards, chiefly grass and crops, through which it runs with great speed, holding up its tail in a folded shape like a common hen's. Other bustards change the shape of their tails in this way too, but not so much as the likh. The fowl-like tail carriage and partiality for the warragoo crops account for the Tamil name *Warragoo kolee* (Warragoo fowl), while it is also called *Khar-titar* (grass-partridge), by the Bheels near Mhow.

The food of this bustard is chiefly grasshoppers, but also centipedes, small lizards, and beetles, including blister-beetles; when these are being consumed its flesh is a viand to be avoided, as mentioned under the heading of the large florican. Ordinarily it is not considered equal to that bird, though, nevertheless, generally in high estimation; but of course with all mixed-feeding birds, the previous diet of the game itself has a good deal to do with the judgment passed on it.

Owing to the wide range of the bird and its nomadic habits, the time of the breeding season varies a good deal, from July to October, according to locality, the northern-breeding birds being later in the year than the southern members of their species. It is a bird which, on account of its unique character no less than its sporting value, deserves careful protection to encourage its increase, and no doubt this could be done better by prohibiting the shooting of hens at any time than by trying to fix close seasons.

The eggs are laid—for no nest but a "scrape" is made—on some little bare patch or among low grass, not over two feet high, no doubt so that the chicks can get about more easily than would be possible in the usual higher cover the old ones frequent. The eggs are broadly oval, rather under two inches long, and speckled with cloudy markings of some shade of brown on a ground of more or less bright greeny drab or brown.

Besides the names above alluded to, this small florican is dignified by some natives with the title of ground peafowl, like its big relative, *Tan-mor* in Mahratta, *Kan-noul* in Canarese, and *Nialanimili*, having this significance. It is also called *Chota charat* or small florican.

Houbara.

Houbara macqueenii. *Tilur*, Punjabi.

The houbara is the characteristic bustard of the semi-desert tracts of North-west India; it is of medium size for a bustard, about two and a half feet long, and has plumage so beautifully assimilated to the sandy soil that it is hard to see at all on the ground, at any rate when crouched flat, which it habitually does when alarmed. On the wing its black-and-white quills show it up conspicuously, and in the hand its long fringe-like black-and-white ruff and the delicate grey on the breast, and the bars of the same tint on the tail, make it conspicuously different from our other bustards. The hen only differs from the cock in being smaller and not quite so fully "furnished" in the matter of head and neck plumage, but the sex difference is only comparative, not absolute as in our other bustards. Cocks weigh about four and hens about three pounds.

This bustard does not breed within Indian limits as far as is known, though it is suspected of doing so in Sind; but it is a well-known winter visitor, sometimes arriving as early as the end of August, but usually at least a month later. After April the birds have generally all departed for their breeding haunts—Persia and the Gulf, Baluchistan and Afghanistan. Outside Sind, Rajputana and the Punjab, houbara are mere stragglers; Hume shot one such in the Meerut district.

In the Western Indian dry plain country the houbara may be found either in the more or less thick but low and scrubby natural cover, or among the cereal crops, so long as these are low. It runs well, and often tries to escape in this way, but towards the time of its departure it appears to feel the heat so much as to be disinclined even to run, let alone fly. When it does rise, its flight is heavy and not long-continued, but it can display considerable wing power when attacked by a hawk. Ridiculously exaggerated as are many of the accounts of "protective colouring," there are some cases in which it really does seem to be a very important asset to the creature possessing it; for not only is it generally agreed that a squatting houbara cannot be picked out by human eyes, but even a falcon flown at one has been seen to settle and walk about in utter bewilderment, looking for the prey that had alighted and adopted the plan of literally "lying low."

Often also the houbara avoids the falcon when dropping into cover, and when hard pressed ejects his excrement, which, as in all bustards, is copious, fluid, and very offensive. He is credited with doing this on purpose, when, on trying to escape by "ringing up" like a heron, he finds the hawk just under him; whether or not intentional, however, the action effectually puts the assailant out of the running, for the filthy discharge so glues its feathers together that it cannot fly well till cleansed, and may drop on the spot. But the falcon chiefly acts as the houbara's foe under the management of man; eagles, which strike at birds on the ground rather than on the wing, are probably the enemies against which the sandy plumage is a disguise.

In thick cover, houbara can be walked up by a line of guns and beaters; where they are to be found in the open, a good way to approach them is to ride round them in diminishing circles on a camel, being ready for them to get up suddenly after they have disappeared by squatting as the approach becomes closer. You can see them all right when standing, in spite of the supposed "counter-shading" effect of the white under-surface of the body. Houbara are usually in parties, sometimes as many as twenty together; they feed chiefly on vegetable food, ber fruits, grewia berries, lemon-grass shoots, and young wheat; now and then beetles and snails are taken, however. They appear to have no sort of call, but the male has a very striking nuptial display; he turns his tail forward, drops his wings, draws his head back and puffs out his neck till the bristling ruff produces a most extraordinary effect. The hen lays, in the usual bustard fashion, on the ground, two or three eggs, elliptical, about two and a half inches long, and stone-colour to olive-brown in tint, with evenly distributed blotches and spots of dark brown and pale purple; some specimens have a green ground. This species of bustard finds its western limit in Mesopotamia as a rule, but it straggles west as well as east, especially to South-eastern Europe. It is found in the highlands of West China, and resides in Afghanistan and Baluchistan all the year round. It goes down in British bird books as Macqueen's bustard, one of our rarities. Somewhere or other it must meet the North African houbara (*Houbara undulata*), which ranges into Armenia, and differs from our bird by having no black tips to the crest feathers, the long breast-plumes white instead of grey, and much coarser black pencilling on the sandy back, this pencilling, common to bustards generally, being particularly delicate in the Eastern houbara. The name *Houbara* is a native one as well as *Tiloor*, which in Sind becomes *Taloor*.

Little Bustard.

Otis tetrax. *Chota tilur*, Hindustani.

" Butterfly houbara " is a common sportsman's name for this smart little bustard, the smallest kind known except the lesser florican ; it is given from the bird's peculiar free-and-easy, go-as-you-please style of flight, often high in the air, and altogether different from that of other bustards. The bird, however, has two distinct styles of flight, for it can get away steadily and swiftly like a partridge. In any case, its white wings make it conspicuous in flight ; on the ground it looks like a bob-tailed hen pheasant, being of about that size and with light brown plumage coarsely mottled with black.

It is only a winter visitor here—and only to the North-west Punjab at that, though stragglers may cross the Indus, and three have been got in Kashmir ; and so we do not see the males in their courting bravery of grey face, white necklaces, and black breast ; in their winter dress they are indistinguishable from hens except in the hand, being merely less coarsely pencilled on the back and less regularly on the breast. There is no constant difference in size, some cocks being smaller than hens, others larger.

October is the month in which the little bustard may be expected to arrive, and most have left India by the end of March ; but it is not much observed or written about, though many are shot or killed with hawks; what is most noticed seems to be the eccentric flight, which renders it impossible to drive the birds. In Europe they have been found to "ring up" and put in about half-an-hour at aerial gymnastics till beaters had gone on, when they resumed their ground and everyday life.

It may be mentioned that the peculiarity of two distinct styles of flight is observable in a very different bird, the beautiful rosy cockatoo or "galah" of Australia (*Cacatua roseicapilla*); this bird, as I was able to observe in specimens they have had loose at the London Zoo on two occasions during the past two years (1912-1913), goes off as heavily as a duck, but when well in the air swings along with the slow, rocking, happy-go-lucky flight of a gull, the very antipodes of the swift flight of the familiar Indian parrakeets.

To return to the butterfly houbara, it frequents low grass and oil-seed crops, and feeds largely on the leaves of the *sarson* mustard, also eating insects and snails; in mustard fields it often lies very close, and is easily walked up and killed. In rising it makes a sharp "pat-pat" with the wings. Its call, "a loud guttural rattling cry," is frequently uttered on the wing; during the breeding season the male calls *trec trec*, and shows off with head drawn back, tail expanded, and half-opened wings; but he is monogamous, and does not yearn after a harem like so many bustards.

The breeding range of the species covers the countries of the Mediterranean basin, and extends eastwards to Central Asia; northwards it ranges to East Prussia and South Russia, where it is one of the characteristic steppe birds, living concealed in the rank vegetation of the region during the summer. In the lush herbage the birds frequent the eggs are very hard to find; the clutch, often in a fairly well-made nest, is large for a bustard, four being laid, which no doubt partly accounts for this species being so common in many localities; but there may be five, three, or two. The eggs are short pointed ovals, olive in ground with a brown or green tinge, and markings of light brown which are often so faint as to be with difficulty distinguishable.

Hume did not think well of the flesh of this bird, which he describes as dark and hard, and rather unpleasantly flavoured;

1/5

OTIS TARDA.

but it is generally spoken well of by European ornithologists. In Baluchistan, where these birds are locally very common, the native name is *Charaz*.

European Great Bustard.

Otis tarda. *Deo-dagh*, Chitral.

Up to date, only four specimens of this great bird have been obtained in Indian limits, all of them hens, and the first as long ago as 1870; in size it about equals the great Indian bustard, sex for sex, but will be easily distinguishable, if met with, on the ground by the absence of the dark cap, head and neck being uniform light grey, and in flight by the white wings, which have only the quills black; the birds keep more together than the large Indian species. Close at hand the coarse barring of black on buff of the upper parts is very different from the finely pencilled dull brown of the great Indian bustard, and the big male, as large as a swan, has in the breeding season long bristly moustaches, but these disappear for the winter, at which season or in spring all Indian specimens have been taken. The bill of this species, like that of the little bustard, being much shorter and more fowl-like than the pigeon-like bill of our Indian well-known kinds, the skull alone would be sufficient evidence of the capture of one.

The first Indian specimen, one of a flock, was got at Mardan, and forty years later two more were killed in the same locality, the weather being very cold. In 1911 two birds occurred, one at Jacobabad in January and one at Chitral in March; all these unfortunate stragglers were young birds and, as above remarked, hens. Birds of this sex are about as big as a grey goose, and indeed there is something goose-like about this species, in its sturdy build, social habits, and fondness for vegetable food; it devours the leaves, ears and seed of a great variety of plants, and, though a great eater, is somewhat of an epicure. Rape is a favourite plant with it, and in some parts of the continent it is classed as a destructive bird. However, it does partake also of insects, worms, and other small animals, and the young are quite insectivorous.

The display of this magnificent game bird is a wonderful

sight. The courting male turns his tail right over on his back and holds it down, as it were, with the crossed tips of his long quills, while the pinion joints of the wing extend outwards and downwards; at the same time the head is thrown back, the neck blown out like a bladder and the whiskers extended on each side. The white feathers on the under-side of the rump and on the wings are fully erected and turned forward, and altogether the bird looks about as unbird-like an object as can be imagined. A specimen well stuffed in this position can be seen in the Natural History Museum, at South Kensington.

The male birds not only show off, but fight furiously for the females. They for their part conceal their eggs with great care. These are generally two, sometimes three, in number, laid in a mere "scrape" or the scantiest of nests; they are about three inches long and olive in hue, tinged with brown or green, variegated with dark brown blotches and spots. The down of the chicks has black spots on a light buff ground. This, the noblest of all bustards, perhaps of all game birds, is also the most widely distributed of the bustard family, ranging all over Central and Southern Europe where conditions are suitable, *i.e.*, in open unenclosed country, cultivated or waste, and east through Asia Minor and Central Asia, to our frontier, while the so-called Dybowski's bustard (*Otis dybowskii*), whose range is from Central Asia to China, is merely a local race, not a fully distinct species. As most people know, it was formerly a well-known though local British bird, but was idiotically allowed to be exterminated; the last flock seen, curiously enough, appeared in the same year as the first Indian specimen. This bustard, in spite of the slow strokes of its wings, flies fast; it is a migrant where winter rigour compels it to wander for food. As human diet it is not so much favoured as it used to be, but its strong smell does not always imply that its meat will be bad eating.

Common Sand-grouse.

Pteroclurus exustus. *Bhat-titar*, Hindustani.

This sandy-coloured, sharp-tailed bird, about the size of a dove and rather like one—indeed, sand-grouse are often, though quite mistakenly, alluded to as "rock-pigeons"—is pretty certain to be encountered by anyone shooting in any dry open district in India; on hills, in heavy cover, and on damp land it is not to be found, and so is generally absent from Bengal, and is not to be looked for either on the Bombay and Malabar coasts nor in Ceylon or Burma, from which all sand-grouse are absent. The dry North-west is naturally its greatest stronghold.

This bird is less handsomely marked than the generality of its allies, which are remarkable for the quiet beauty of their plumage. The sandy hue of the cock is relieved, however, by yellow on the throat and face, and chocolate on the belly; the hen is barred with black on the buff upper-parts and has the dark brown abdomen lightened by buff barring.

These sand-grouse begin feeding at daybreak, frequenting stubble fields and weedy fallow land, where they procure the seeds of the weeds; and they also eat millet and pulse, as well as grass seeds, but insects are very rarely eaten. Hume only notes two cases of this, the insects in one case being ants, in the other beetles. In both specimens seeds were present as well.

Between 8 and 10 o'clock in the morning they go to the nearest water to drink. They have sometimes to fly a long distance for this purpose, but this does not matter to them, as they are fine and swift fliers, and often fly very high; they call continually when on the wing, their note being a double cluck. After the morning feed they seek another feeding-ground of a more open character, such as a ploughed field or sandy plain; here, after a slight lunch, they lie down to take a nap during the heat of the day under some clod or other shade, getting up again for tea in another field, and gradually drifting off later on for the evening drink from 4 to 6 o'clock. The drinking time varies with the season, but, whenever it is, their enemies are liable to interfere with this simple routine of existence, since it is the custom to lie up for them at the drinking place. Owing to their speed they afford good sport, but, as Hume says, it is cruel to wait for them at both drinking times where water is scarce. They cannot apparently go very long without a drink, if they miss one in the morning; but when brutally scared off again they will leave the neighbourhood, even if they have eggs close by.

This may be the case in suitable localities at almost any time of year, for this species is a resident and has a very extended breeding season, though in the North-west April to June are the usual months. Both sexes care for the eggs and young, the cocks sitting by night and the hens by day, and although the chicks pick for themselves, water is brought for them by the male parent, who soaks his breast feathers in it and lets the chicks suck it off. This no doubt accounts for Hume seeing the birds washing, as he thought, at their drinking places; the habit of thus watering the young, which has been discovered in recent years by people who have bred sand-grouse in aviaries, was not then known. In the ordinary way, dusting rather than washing is the sand-grouse custom, so when birds are seen wetting their plumage at the drinking places it may be suspected that they have young somewhere near, and no shooting should be done.

The eggs are laid in a scrape on the ground, occasionally with a scanty lining, and are usually two, rarely more or fewer. They are of long shape and blunt at both ends—this peculiar form of egg being characteristic of the eggs of the sand-grouse family—and the ground-colour varies, being greyish-white, pinkish stone-colour, cream, or olive-brown. The spots are dark brown and dull mauve, their intensity and amount varying very much.

These sand-grouse sleep at night on the ground, like all the group, selecting some open place, and Hume remarks on their extreme watchfulness. They pack closely at such times, not scattering as they do during the noonday siesta, and no doubt find that "many heads are better than one" in the matter of keeping a look-out; at any rate, they seem never to fall into the clutches of the ordinary four-footed vermin, though native fowlers catch them sometimes.

$\frac{1}{2}$ PTEROCLES SENEGALUS

These birds weigh about eight ounces, the cocks running heavier; they also have the long feathers in the tail about an inch longer than the hens.

Although a resident with us, this sand-grouse is not confined to India, inhabiting also central and south-west Asia and a large part of Africa; it is, in fact, truly the common sand-grouse, having the widest range in the family. Being so noticeable and well known a bird in India, it has many names—*Bukht-titar*, *Kumar-tit* and *Kuhar*—as well as the one above given in Hindustani, *Pakorade* or *Pokundi* in Marathi, *Jam polanka* in Telugu, *Kal-kondari* in Tamil, and *Kal-gowjal haki* in Canarese, while in Sind the names are *Butabur* and *Batobun*, and among the Bheels *Popandi* is used.

Spotted Sand-grouse.

Pteroclurus senegallus. *Nandu katinga*, Sindi.

It is not often that a bird is named from a peculiarity in the hen's plumage alone, but that is the case with the present one, in which only the hen is spotted, the spots being small round black ones on a buff ground, and taking the place of the more or less transverse black pencilling usual in female sand-grouse. Except for this, she is very like the common sand-grouse, and has a similar pin tail; the cock has a corresponding likeness to that of the common *Bhut-titar*, but is easily distinguishable by his grey eye-streaks and chocolate mottling on the wings.

This species, though numerous where it occurs, is very local, being a regular visitor only to Sind and Jeysulmere; it also occurs near the Runn of Cutch and in the Shahpur district in the Punjab. Although for the most part a winter visitor, some individuals undoubtedly breed in Sind, but the main haunts of the species are in Northern Africa, though it is also found in South-west Asia.

The spotted sand-grouse assembles in flocks up to fifty in number and has a strong preference for the desert, though it comes to the edge of cultivation for food. The food includes more insects than is usually the case with sand-grouse, though seeds form the main portion of it; but Hume was no doubt right in thinking that to this more insectivorous tendency the greater succulence of the present bird's flesh is due, though he had a poor opinion of any sort of sand-grouse as food, except when "baked in a ball of clay, gipsy fashion." I should think that wrapped in a thin rasher of bacon, civilized fashion, they might go just as well.

The note of this specics is characteristic, being, according to Hume and Dresser, a gurgling sound like "quiddle, quiddle, quiddle"; but James, quoted by Hume, says it is very like that of the common sand-grouse, though he admits it is less harsh, and easily distinguishable.

It is not so wary as the black-bellied sand-grouse, and the cocks are less quarrelsome than in that species. The eggs are laid in March and April, and number two or three; one extracted from a hen shot in Upper Sind is described as pale yellowish stone colour, spotted with olive-brown and grey. The weight of the bird is about nine ounces or more in cocks and less in hens.

Large Pintailed Sand-grouse.

Pteroclurus alchata. *El Guett'ha,* Arabic.

Nobody seems to know any special native name for this sand-grouse, which is curious, seeing that for distinctness and beauty of plumage it stands apart, and is one of our three species which are as large as pigeons. From the other two, the Tibetan and the black-bellied, it is easily distinguished; its belly is white, not black, as in the latter, and though the Tibetan sand-grouse has also a white belly, its toes, feathered as well as the legs, distinguish it. Moreover, the present bird has, above the white belly, a black band clearly marking off the buff breast. These points, and the long pin feathers in the tail, are common to both sexes, but otherwise they differ; the hen has the usual black-pencilled buff upper plumage of hen sand-grouse, and the cock's upper plumage is a beautiful mixture of subdued olive-green and soft yellow, the former colour predominating. He has a black throat and a black necklace a little way below this, defining the

Pl.II.

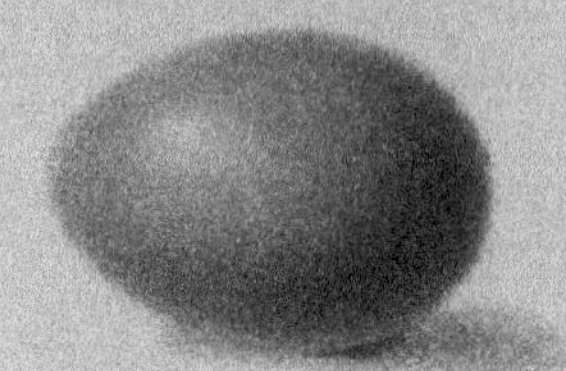

Porzana bailloni.

Pterocles fasciatus.

Megapodius nicobariensis

Rhynchæa bengalensis.

Hypotænidia striata.

R. W. Stott. Del.

F. Waller Lith 18, Hatton Garden, London.

top of the buff breast from the lighter buff of the neck. The hen has also a necklace here, but it is double in her. Both have a beautiful variegated patch on the wing near the pinion-joint, formed by white-edged feathers, their ground-colour being chocolate in the cock and black in the hen.

It may be that the very local occurrence of this exquisitely beautiful, and, where it occurs, very abundant bird, have prevented it getting a distinct appellation in our Indian tongues; it is only at all common in Northern and Central Sind and the Punjab, and only a winter visitor there, but it has strayed to Delhi and been obtained in Rajputana. It leaves for the north early in April at latest; its range is very wide, and it must be one of the most abundant of the family, being found not only in South-western and Central Asia, but being represented in Southern Europe and North Africa by what can hardly be called a distinct species, though it can just be separated as a local race, the so-called *P. pyrenaicus*.

When with us it is in enormous flocks; they were, says Hume, in tens of thousands on a vast plain some miles from Hoti Mardan, the only place where he had had a chance of observing them; here they only frequented the bare and fallow land, though cultivated ground was at hand, and were very wary, requiring to be stalked by way of nullahs or ravines. Their flight seemed to him even more powerful than that of other sand-grouse, and the note, uttered freely either on the ground or on the wing, was, although rather like that of the black-bellied species, still quite distinct from that of any other. Dresser renders it as *kaat kaat ka*, and Blanford calls it a loud clanging cry. The food is leaves, seeds, small pulse and grain, and plenty of gravel is taken.

This is one of the species the male of which has been actually observed to soak its breast to water its young, as recorded by Mr. E. G. Meade-Waldo, the first to discover this unique habit in 1895. He says, in the *Avicultural Magazine* (1905-1906), "I have had the good fortune to see the males of *Pterocles arenarius*, the black-breasted sand-grouse, and *Pterocles alchatus*, the greater pintailed sand-grouse, getting water for their young in the wild state, but, had I not seen it administered in confinement, would have considered them to have been demented birds trying to dust in mud and water, when unlimited dusting ground surrounded them on every side." He appears not to have seen the actual act of soaking by the present species, but saw the cocks pass over with their white breasts soaked in mud and water, and he has often bred this bird (the western form) in confinement, his old hen, the mother of many broods, having died at the advanced age of nearly 20 years.

Black-bellied Sand-grouse.

Pterocles arenarius. *Burra bhatta*, Hindustani.

The black belly which gives this fine sand-grouse its name will also distinguish it from all our other species ; it is one of our three large kinds, being as big as any ordinary pigeon. The hen has the usual buff black-speckled plumage above, though black below like her mate ; the cock has the head and breast grey, the throat chestnut, and the upper parts elsewhere variegated with grey and buffy-yellow. There are no long points to the centre tail feathers as in our other large species.

Although only a winter visitant to India, this sand-grouse is probably the most numerous after the common kind, since it arrives in enormous numbers. Hume, for instance, says "Driving in November, 1867, the last stage into Fázilka from Ferozepore, parallel to and on the average about two miles distant from the Sutlej, over a hundred flocks, or parties of from four or five to close upon one hundred each, flew over us during our fifteen miles drive; they were all going to the river to drink or returning thence. Necessarily we can only have seen an exceedingly small fraction of the total number that that morning crossed that little stretch of road."

But, although thus abundant locally, the black-bellied sand-grouse is not widely distributed in India during its stay there, but is confined to the North-west Provinces, and in most years only really abundant, says Hume, in northern and western Rajputana and the Punjab west of Umballa ; also, being averse to heat, it comes rather late, about the middle of October and leaves as a rule at the end of February. It has a wide range outside India, throughout South-west Asia and Northern Africa and to Spain and even to the Canary Islands; in these parts of its range it breeds and has been found doing so even in Afghanistan, the time of year being May. It is also found in Turkestan.

This species especially affects wide plains in sandy country, and if water is reasonably near, such localities in the North-west are a pretty safe find for them during the cold months they stay with us. Large numbers are bagged at their drinking places, three guns at three different tanks having once bagged fifty-four brace in three hours in a morning, as reported to Hume by his native friend Khan Nizam-ud-din, who was one of the guns in question.

They generally prefer ploughed land to stubble, and when washing in the early morning pack very closely, though they scatter more during the mid-day siesta, and then lie on either side alternately to enjoy the heat, with one wing slightly extended. They are inclined to be wary, especially when they have been much shot at. The note is very much like that of the common sand-grouse and the native names are generally the same, though *Banchur* is used at Peshawar for the big bird.

Like the common sand-grouse also, most of their food is seeds with the addition of bits of herbage; insects are also eaten occasionally, though hardly ever in Hume's experience. They straggle a good deal when feeding, and towards the time of their departure the males skirmish a good deal without actual stand-up fighting. The weight is about a pound, the cocks being inclined to run over this weight and the hens to be lighter.

Sand-grouse of all kinds strike one at once as being especially suited to the table on account of the huge breast muscles which move their long wings in their swift, powerful flight and make up the greater part of the body; but in practice they are rather apt to be dry when brought to table, though very freely used where they can be got.

Coroneted Sand-grouse.

Pterocles coronatus.

This, like the common and spotted sand-grouse, is a sandy-coloured bird of the size of a dove, but has not the long feathers in the tail; the cock has a grey stripe on the side of the head, and some chocolate on the wings, but can be easily distinguished from the spotted sand-grouse and common sand-grouse cocks by his small black bib and black streaks on the forehead and face. The hen is less easy to distinguish, but may be known from the hen spotted sand-grouse by the black markings on her buff plumage being crescents, not spots, while the absence of the long points to the tail will distinguish her from the hen common sand-grouse.

Like the Lichtenstein's sand-grouse, this is merely a frontier bird of doubtful status; it is uncommon and usually only occurs in Sind, does not go east of the Indus even there, and is only suspected of breeding in the country, being usually a winter visitor. Its present range is from North-east Africa, through Arabia, South Persia and Baluchistan, to our frontiers. It breeds as near as Afghanistan, eggs being found about May or June; they are pale in colour, the spots being scanty and pale brown, while the ground colour is greyish white.

There is very little on record about this bird; Heuglin says it is just like the spotted sand-grouse—which, in spite of being "pintailed," appears to be its nearest ally in voice and habits. In the southern parts of the Sahara according to Tristram, it takes the place of the black-bellied sand-grouse, along with the spotted species. He only found it in small parties of four or five (no doubt families), but he attributed this to the extreme scarcity of vegetation in its arid haunts. In Baluchistan, Blanford considered it commoner than the spotted sand-grouse.

Mr. Whitaker gives some useful notes on it in his "Birds of Tunisia," where he found it abundant, though local, in the districts south of the Atlas. He found it coming in flocks of from ten to fifty birds to drink at the water-holes at Oglet

Alima, from 7 to 10 a.m. They did not come back that evening, but turned up again next morning. They flew high and very strongly, with a loud clucking note somewhat like that of a fowl, audible even when the birds themselves were so high as hardly to be seen. They were shy and required a hard blow to bring them down, but were "excellent eating, not at all dry or tasteless, the breast having dark and light meat the same as black game." He never found anything in their stomachs but seed and other vegetable matter.

Painted Sand-grouse.

Pterocles fasciatus. *Pahari bhat-titar*, Hindustani.

This is the smallest of our well-known sand-grouse, and of remarkable beauty of plumage, at any rate in the case of the male, whose buff ground colour is diversified above by broad and close-set chocolate bands, about as wide as the interspaces between them. The wings have a few white bands, and the head two bands each of black-and-white; the breast is unbanded, but below this the plumage is coloured with black and buff in narrow equal bands. The orange bill and yellow eye-lids are also characteristic colour-points of this species. There are no long "pin" feathers in the tail.

The hen is of a more ordinary type, barred with buff and black nearly all over, the head being spotted with black, not barred.

This sand-grouse is less short-legged and squatty than the others, and not so long in the wings; it approaches the partridge type more, in fact, and in accordance with this structure it is found to run much more quickly and freely than other sand-grouse, so that it might almost at times be taken for a partridge. It also frequents rather different localities from other sand-grouse, for though, like the rest of them, liking dry soil, it is found in places where there is a good deal of bush and even tree cover. Nor is it typically a bird of the plains, for it especially affects hills and ravines, and likes to frequent the mounds left when a jungle village has been deserted. In the rains it deserts its bushy haunts and lives in the open like the common sand-grouse. Where such ground as it likes is to be found it is widely spread, but not by any means universally; its chief stronghold is in Guzerat, Cutch, Rajputana, the North-west Provinces and the Simaliks. On the Malabar and Bombay coasts and the Ganges delta and the Carnatic lowlands it is not to be found, nor does it pass westward of the Indus. It is not nearly so gregarious as other sand-grouse, being usually found in pairs or even singly, a flock of ten being exceptional. Owing to its lying very close, at any rate in the day-time, it is seldom seen until startled, when the birds fly strongly and fast, but are not difficult to kill. They do not fly very far at a time, and do not go far from home except in search of water, which they visit much earlier and later than other species, not taking their evening drink till dusk or even after dark, and drinking again before sunrise. They have also often been seen feeding and flying about at night, and observations on specimens in captivity in England have shown them so very lethargic by day that it is probable that they are really night—rather than day—birds, which would no doubt account for the fact of their lying so much closer than other species.

Their note is as characteristic as their other traits; although they call on the wing like other sand-grouse, the note is described by Hume as a "chuckling chirp."

The nest of this bird is, however, the usual sand-grouse "scrape" on the ground, with the scantiest of lining, if any; the eggs, which are usually three in number, but sometimes two or even four, are of the typical long shape, but of a very distinctive colour, salmon-pink, marked with brownish-red and dull purple; they are much like those of some nightjars.

The weight of these birds is between six and seven ounces, the male exceeding the female but very little. In South India it is known as *Handeri*, the Tamil word being *Sonda polanka*.

$\frac{9}{16}$

PTEROCLES LICHTENSTEINI.

Close-barred Sand-grouse.

Pterocles lichtensteini.

The close, narrow, transverse black barring on the upper parts and breast of the male of this bird at once distinguish it from that of the painted sand-grouse, to which it is, nevertheless, very similar and closely related; the hens are, as one would expect, much more alike, but the presence in the present bird of only fourteen instead of sixteen tail-feathers, and the absence of the bars on the "stockings," or leg-feathering—the painted, alone among our sand-grouse, going in for barred hose—are certain distinctions between them; moreover, on the abdomen of the painted sand-grouse, in both sexes, the black predominates over the white, while the reverse is the case with Lichtenstein's.

The close-barred sand-grouse is only a frontier bird with us, visiting Sind only, and not penetrating further east than the Indus; its usual home is North-east Africa, Arabia, and Baluchistan, and it is resident in those countries. It may, perhaps, be found to breed in Sind, although generally ranked only as a cold-weather bird, and one of very irregular occurrence at that.

When found, it is, like the painted sand-grouse, in pairs, or small parties, frequenting scrub and rocky ground near cultivation. It does not fly far at a time, and lies well, but is not always easy to hit, especially in the dusk, for, again like its ally, it has decidedly nocturnal tendencies, and comes to water very late in the evening and very early in the morning.

"On moonlight nights," says von Heuglin, speaking of its habits in North-east Africa, "these birds never roost at all, and there is really no end to the clapping and striking of wings and the whistling and croaking of these noisy fowl as they straggle about on the ground, especially in the neighbourhood of the desert springs, with lowered pinions and upturned and outspread tails." They feed, however, he says, in the forenoon and again towards evening, frequenting fields of maize, cotton, and indigo, threshing-floors, caravan-roads, and weedy valleys. The note is quite different from that of the coroneted and common sand-grouse, and resembles a sharp whistle through the fingers, and is "deafening" when the small parties pack in numbers at their drinking-places. As he speaks of them as in enormous multitudes in some places, they must be far more numerous in their African haunts than they ever seem to be in India, and far more so than the painted sand-grouse ever is, for that matter; though, no doubt, scarcity of watering-places causes a great deal of concentration.

The eggs, according to Heuglin, although of the usual long shape of sand-grouse eggs, are "much the colour of dirty and faded peewit's eggs." The highland slopes with a thin covering of scrub were the breeding-grounds, and the time of breeding the beginning of the rains.

Tibetan Sand-grouse.

Syrrhaptes tibetanus. *Kuk*, Ladakhi.

The Tibetan sand-grouse is at once distinguishable from all other Indian game-birds by having the legs and toes—which by the way are excessively short—completely feathered to the claws; there is no hind-toe at all. It is larger than any of the other sand-grouse, and duller in colour, its sandy hue being only relieved by dull-orange neck and cheeks, a white belly, and black pinion quills; the tail has long "pin feathers" in the middle. The hen's black pencilling on breast and back will easily distinguish her, for though the cock is also pencilled above, the markings are very fine and not conspicuous.

The Ladakhi name of this bird is evidently derived from its characteristic cry, which, however, is in two syllables: another local name is *Kaling*. The only other place, besides Ladakh, in which this species occurs in our Empire is the valley of the upper Sutlej; its real home is the "Roof of the World," the high steppes of Tibet and the Pamir Plateau, extending to Koko-Nor; but it may of course be expected frequently to stray over our frontier at high levels. Its haunts are barren and desolate places, but it manages to find sufficient food to exist upon in the shape of grass-seeds, shoots and berries, in search of

½ SYRRHAPTES TIBETANUS.

which it shows more activity in getting about than one would expect in such a very short-legged bird—it is quite the squattiest in the game-list. Flocks of hundreds often occur, but in summer these break up into little groups. They are very hard to see when basking in the sun at midday, owing to their plumage being so like the sand, and make a prodigious noise as they get up suddenly, and rather surprisingly; for at such times they lie very close, and in spite of their fast and powerful flight, do not go far, and may be marked down and flushed again and again.

In the mornings and evenings they are apt to be much more shy, and to take alarm at a hundred yards' distance. The drinking-times—two in the twenty-four hours, as usual with sand-grouse—are in the very early morning and quite at dusk. The birds are generally near water, and will drink brackish if they cannot get fresh. They are noisy birds when on the move, and their characteristic double cluck can be heard at night as well as by day.

The eggs have never been taken within British limits, and in fact till within quite recent years were not known at all. However, there are in the British Museum collection a couple taken on the Pamir, and presented by Mr. St. George Littledale, which are described in the Museum Catalogue of Birds' Eggs as follows :—

"The eggs of the Tibetan three-toed sand-grouse in the collection are of a pale creamy-buff colour. Both the shell-markings and the surface-markings are small, and the latter consist entirely of spots of dull reddish-brown evenly distributed over the whole shell. Two examples measure respectively 1·9 × 1·37 : 2 × 1·33."

Peacock.

Pavo cristatus. *Mor*, Hindustani.

India and Ceylon are the native and only homes of the common peacock, a bird so well known in domestication everywhere that there would really be no need to describe it were it not for the sake of pointing out some sexual, varietal, and specific distinctions.

With regard to the first, everyone knows that the peahen is a plain brown bird without the cock's long train, but it should be noted that she has a similar crest, a good deal of green on the neck, and the under-parts dirty-white. The young cock in his first year is very like her, but can be known at once, if put on the wing, by the bright cinnamon pinion-quills, the first sign of the masculine plumage to be fully developed ; though even at this early age the neck is more glossy and bluer than the hen's, and the pencilling of the wings is indicated.

In the next year the full rich blue of the neck and the distinct chequering of black and buff on the wings are developed, together with the scale-like golden feathers between the shoulders. The train, however, is foreshadowed only by some beautiful bronze-green feathers overhanging most of the tail, but without eyes or fringes, and quite short. In the next year the full train is assumed, and no further change takes place, except that for some years it lengthens a little. The age of the birds can therefore not be judged for more than a few years, but there is good reason to believe that it may reach a century in captivity, though such a long life is hardly within the expectation of the wild peacock, with all the risks he has to run. Chicks have the brown down, streaked with darker, so commonly found among the true game-birds; in the brown chicken-feather they have the long narrow crest of the other species, the Burmese or Javan peafowl. They begin to show off when as small as a partridge, or smaller.

In the so-called *Pavo nigripennis*, the black-winged or Japan peafowl, which has not been found wild in Japan or anywhere else—with the exception of a hen once shot in the Doon and seen in the skin by Hume—the cock has those parts of the wings which are speckled black and cream in the ordinary form, black glossed at the edges with blue and green, and the thighs black instead of the usual drab. The corresponding hen is white with black tail, a considerable but variable amount of black streaking and peppering above, cinnamon crown and pinion-quills, and white instead of dark legs, as indeed the black-winged cock has. In the down, birds of this variety are primrose-yellow, developing buffy wing-feathers, and ultimately a chicken-feather of cream colour barred above with black, from which the cocks by degrees become darker and the hens lighter, as described above. I have gone into so much detail about this bird because it is positively known to arise as a "sport" from ordinary domestic peafowl, and to produce ordinary birds when crossed with the white variety so often imported to India from Europe as a curiosity. There is, therefore, no doubt that it is not distinct from the ordinary peafowl; but at the same time it is a most interesting form, as it is distinct in all stages and rarely fails to reproduce its kind, so that any information about its occurrence in the wild state would be very interesting. The "uniformly dirty-yellow" hens seen wild by Sanderson were presumably a buff form, or they might have been young hens of this type, but as he says nothing about markings the former is more likely, and as I have heard of a cock of the domestic race which was described to me as exactly of the colour of a new copper coin, it is evident that a buff variation is possible for both sexes. Pied birds also are well known in domestication, but with the exception of the production of these colour varieties the peacock has not altered at all since its first introduction into Europe following on Alexander the Great's invasion of India, so that twenty centuries of domestication in an alien climate have not affected its plumage any more than they have eliminated the less elaborate but even more conspicuous black-and-red hues of its companion, the jungle-fowl.

Like that bird, the peacock is essentially a lowlander, not ascending the hills into a temperate climate, though in the hills of the south of India it goes higher than in the Himalayas. It likes tree cover near water and cultivation, and where such conditions occur may be found almost anywhere in India and Ceylon; but, as Hume says, there may be too much water, cover, and cultivation to suit it, and so it is local. It is, however, often found in places which do not seem by any means ideal, such as the sandy semi-desert parts of the North-west. Here it is protected by its sanctity in Hindu eyes, and indeed everywhere where Hinduism is dominant; in native states no one may shoot it, and in any case it is always well to ascertain the state of local religious feeling before firing at peafowl, or the consequences may easily be serious.

No doubt in many out-of-the-way and naturally unsuitable places these birds have been artificially introduced, and where rigidly protected they are half tame; but where they have to take their chance few birds are wilder and more wary, and they are most untiring runners, only taking to wing as a last resort. The flight is quite different from that of a pheasant, the wings moving with comparatively leisurely flaps and no sailing intervals; however, it is much faster than it looks, and the birds, if positively forced, can rocket like any pheasant; but the old cocks, whose long trains are so much dead weight, cannot fly very far at a time, and have been even run down by persistent chasing during the hot weather. It is probable that any shifting of ground that the peafowl have to make is done mostly on foot, as used to be the case with wild turkeys in America in the days of their abundance; these birds, by the way, having also limited powers of flight, often fell into rivers and had to swim ashore, and I have seen in England a young peacock reduced to the same extremity by having tried to fly across a stream with clipped wings, save himself similarly, swimming as readily as a moorhen.

The scream of the peacock is very well known, but the ordinary call-note is less familiar; it sounds like anyone trying to pronounce the bird's Latin name *Pavo* through a trumpet, and is often used as an alarm-call. Being essentially birds of tree-jungle, pea-fowl naturally roost on trees, and high trees at that; but they do not mount to the top, but settle down on the lower boughs. They are late in roosting in the wild state, and sometimes in domestication, though I have commonly observed them going to bed quite early. Yet they are wary at night—at any rate an escaped hen in Covent Garden defied nocturnal surprises

$\frac{1}{6}$

PAVO MUTICUS

for several months ; but they can be shot on the roost in the wild state, though only need of food ought to drive anyone to this. The buff eggs, by the way, about half a dozen of which are laid, generally in the rains and on the ground, are most excellent.

It must be admitted, however, that peafowl are not by any means friends to the farmer and forester, as they are destructive to grain, herbage, flowers, and buds ; most of their food is, in fact, vegetable, but they also, to their credit, consume various insects and other vermin, including young snakes. They are as good to eat as turkeys, if yearlings are taken, and a yearling cock can always be picked out, as I said above, by his cinnamon quills. Young hens have slightly redder quills than old ones, and are a little pencilled on the feathers over the tail. As the cock is three years old by the time he is in full colour, one can only expect him to be tough, as any ordinary rooster would be ; but of course he is good for soup.

The peacock of course has names in all the native languages, and sometimes the cock and hen have different ones ; thus in Uriya *Manja* is the cock and *Mania* the hen ; in Mahratta there is a still greater difference, the cock being *Taus*, very close to the Greek *Taôs*, and the hen *Landuri ;* the Assamese *Moir* comes very near the Hindustani name, which has also the variant *Manjar ; Nowl* is the name in Canarese, *Nimili* in Jeluga, *Myl* in Tamil ; the Lepcha word is *Morg-yung* and the Bhotanese *Mabja*, while the Garos use *Dode*, and the Nepalese *Monara*.

Burmese Peafowl.

Pavo muticus. *Daung*, Burmese.

This Eastern species of peafowl, the only relative and rival of our old friend which figures on a small scale below it on the plate, is also well known in its way from Japanese art, in which it is the only form of peafowl represented ; many people must have noticed that the Japanese artist's peafowl have long narrow crests and neck-feathers as clearly defined as the scales of a fish ; and these points are indeed characteristic of both sexes of the Burmese bird, as well as the predominant green colour of the neck and the plumage generally, on account of which it is sometimes called the green peacock. The long lance-head shaped crest has given the bird one of its scientific names, *Pavo spicifer*, and the dealers at home often call it the "specifer peacock." In this species the bare skin of the face is much more extended than in the Indian peacock, and is richly coloured with orange-yellow on the jaws and cheeks and mauve-blue round the eyes. The train is very like that of the common peacock, but the wings are black in both sexes, and both have cinnamon pinion-quills. In fact, fine hens of this species are almost exactly like cocks, except of course for having no train; and they likewise lack the scale-like shoulder patch, which is greener and less gold in this species than in the other, but they have dark under-parts like the cocks. Yearling cocks already have the green short tail-coverts which the ordinary bird does not get till the second year, and are so like old hens that the only reliable distinction is the colour of the little patch of feathers that breaks the bare facial skin between bill and eye; this is rusty brown even in the best hens, and deep glossy green in any cock. A similar difference may be observed in this patch in common peafowl. As hens of the Burmese bird have spurs as a regular thing, they are of no use as a sex distinction.

In the second year the Burmese pea cockerel assumes the scaly scapular patch and an especial mark of masculinity, a lovely blue patch near the pinion-joint of the wing, an area which is always green in the hen: in the third he gets his full train, so his development is really much like that of his Western cousin in point of stages, though he starts with an advantage.

The note is, however, strikingly different in this bird, being six-syllabled and very subdued and unobtrusive. The bird itself, however, is not by any means so in captivity, for he is extremely spiteful and a most dangerous bird to have about where there are children and infirm people, while his unexpected attacks are not pleasant for anyone. Yet he is quite susceptible of attachment to individuals, and the young birds and hens are charmingly tame. The cock also chiefly shows his fierce temper when in possession of his full train, showing a curious analogy to deer, which are chiefly dangerous when possessing their horns.

The range of this peafowl begins where that of the common

bird ends; it has been recorded from one locality in Cachar, where, however, the other is the ordinary species, as in Assam; it is the only peafowl found in Burma and Malaysia, and ranges eastwards to Java. It must have been taken to Japan many centuries ago, for it was first described by Aldrovandi in the 16th century, from a drawing sent by the Japanese Emperor to the Pope. In Europe it is rare in captivity, and not much is on record about it in the wild state. It is not nearly so common in most places as is the Indian peafowl, being only really abundant in our limits in Upper Burma, and occurring generally in isolated colonies a long distance apart. The general habits appear to be similar to those of the common peafowl, though it is much wilder and less sociable, but there are no doubt other differences in detail. Wallace, speaking of the bird in Java, says it flies over high trees with ease, and an officer I met told me that it could be seen in the evening flighting up the rivers in Burma; this looks as if it flew more freely and readily than the Indian species, and it certainly has longer wings, the pinion-quills showing their tips outside the others in the closed wing. Its remarkably slim and long-legged build is also noticeable; in fact, it is as stilty as many waders, and I have seen hens at the London Zoo, when kept along with cranes, wading and standing in a small pond in cold as well as hot weather, though their mate, a very chilly bird, would not do so. Tickell also says that these birds, as well as jungle-fowl in Burma, especially affect islets in rivers in the evening, scratching in the sand at the margin and roosting safe from vermin. Possibly they wade about also; the domestic fowl in India is certainly a great wader in suburban ditches in Calcutta, or was in my time, when they seemed nearly as aquatic as rails, wading right up to their hocks.

The display of this peafowl is similar to that of the common species, but the wings are brought farther forward so as to brush the legs, and owing to the length of these and the comparative skimpiness of the train, the "nautch" is less grand and imposing. As this green peafowl has occurred in the territory of the other species, hybrids between the two might occasionally occur, so it is worth while to mention the points of some which were bred a few years back in the London Zoo between a Javan hen and a black-winged common peacock. Hens and yearling cocks (the latter were sent away soon after their first year) were much alike and had dark brown plumage, pencilled with buff above, and with no white on the lower parts; the quills were cinnamon in all, and the upper tail-coverts bronzed. So far they resembled more their Javan mother, but the crest was that of the common peafowl, and they had the face-skin equally limited in extent, and nearly as white, but with a vivid orange patch under the ear. The colour of the neck, however, was of a rich glossy emerald, like a mallard's, differing much from the bronze-green and purple of the Javan birds and more resembling that of the nape of the common peahen, though covering the entire neck. A trio of these birds are said to have reproduced again in the grounds of the well-known Dutch aviculturist, Mynheer Blaauw. The eggs of this species, which are laid during the rains, resemble those of common peafowl, and the chicks also appear to be similar.

Grey Peacock Pheasant.

Polyplectrum chinquis. *Deyodahuk*, Assamese.

This beautiful bird's grey plumage, spangled with metallic spots of purple-green, is quite sufficient distinction from any other bird met with in his haunts, which are the lower elevations of the hills, or localities thereto adjacent, from Sikkim to Burma. The hen bird is sufficiently like him in general appearance to be recognizable at once, but she is not only much less in size, weighing only a pound at most, while he may reach one and three-quarters, but also, as might be expected, duller in plumage, especially in having the glittering spangles replaced by sombre black, except at the tip of the tail. Young birds have the tail transversely barred with a light colour at first.

These pea-pheasants, as Jerdon well calls them, are in some ways very like pigmy peafowl; they have the same level-backed carriage, light build, and dainty gliding gait, and the cock in full display, when he stoops and spreads his bejewelled tail before the hen, is very peacock-like; but his display differs in detail in the fact that the wings, being also ornamented, are set out on each side of the tail and enhance the effect, and are only half opened, while the peacock keeps his wings behind his train and even the true tail at the back, and shuffles them with the chestnut flight feathers showing. At the London Zoo it has been observed that the male peacock-pheasant, when about to display, allures the hen by offering her a bit of food and then takes advantage of her proximity to show off, a very intelligent-looking action. Like the tragopan, he has a sideway as well as a frontal show, slanting himself, as it were, so as to show all his spots on one side, and this was for long thought to be his only pose.

In time of courtship his hairy-looking crest, which is always longer than the hen's and is chronically on end, turns right forward over his beak, even when he is not otherwise displaying. No doubt he fights with his rivals, and his legs are often armed with several spurs apiece, but the number is very variable, and some time ago I noted in three males at the Zoo, all imported birds and several years old, that all differed in this point, one indeed having no spurs at all and another only one.

It looks, therefore, as if the idea current among the Kookies, that a new spur grows every year, is incorrect, and that the number of spurs is purely an individual point. The morning and evening call of the cock, which begins with the year, and is uttered at half-minute intervals, often for an hour or more at a time, from a perch on a tree or stump, is described as "something like a laugh"; it certainly is in several syllables, but the laugh is a very harsh one, and I have noted it as a barking cackle. It is deceptive as to distance, and yet furnishes the best means of stalking the bird, which is not at all easy to get at by any sportsmanlike means. It keeps closely to cover, especially bamboos and low trees; only if it can be forced to "tree" by hunting it with noisy curs, it may provide a pot shot. Natives often snare it, and Davison once had a very curious experience in getting specimens in this way in Tenasserim, where he found the bird very common. "I found," he says, as quoted by Hume, "three holes of the porcupine rat (of which I got two specimens) communicating with one another; the entrance to one of these holes was nearly three feet in diameter and some four feet in depth, decreasing, as the hole deepened horizontally into the hill side, to about eight inches. I set a slip noose with a springer in the small part of the holes. On looking next morning, instead of, as I expected, finding the rat, there were only a number of feathers of the male of this species. I set the trap again, and that evening got nothing; next morning I found a hen hanging by her legs in the trap." Here were evidently a pair in the habit of going to ground, a custom which needs investigation. The birds, by the way, generally do pair, and seldom more than four are seen together, such parties being probably families, since in captivity only two eggs are laid at a time. The birds begin breeding about May, retreating to the densest jungles at this time, and ceasing to call till the autumn; the cream-coloured eggs, about two inches long, are laid on the ground under a bush, and the young when hatched run close behind their mother, completely hidden from view by her long broad tail, which she expands to cover them. They are dark chocolate in colour, without stripes, and their slim black legs are noticeable and characteristic, the old birds being particularly fine-limbed, and slaty-black in the colour of these parts.

This peacock-pheasant is excellent eating; it feeds on insects, snails, seeds, and especially on certain red berries which are used by the Kookies as bait for their springes. Fortunately, these catch more cocks than hens, but poaching tricks of this sort ought to be made illegal everywhere; there are plenty of vermin that want killing down everywhere in India, and the destructive energies of natives should be directed on these rather than game-birds. The beautiful plumage of this bird is very suitable for decoration, and if protected during the breeding season it might well be made to supply this.

The tea-garden coolies recognize the affinity of this bird to the peacock by calling it one—*Paisa-walla Majur*. In North-eastern Cachar they are called *Mohr*. *Munnowar* is an Assamese name as well as that given above, of which the Garo *Deo-dirrik* is obviously a variant; while in Tenasserim the name is *Shway dong* and in Arrakan and Pegu *Doung-kulla*. Outside our limits this bird is found in Siam.

Malayan Peacock-pheasant.

Polyplectrum bicalcaratum.

The Malay peacock-pheasant has the same grizzled style of plumage, ornamented with eye-spots, as the better known grey species, but is brown, not grey, and darker in tint, the minute speckling which produces the grizzled effect being black on a ground of pale brown. In addition to this difference, which characterizes both sexes, the cock has a purple or green glossed crest and a red face, while in the hen the eye-spots are much better developed, especially on the tail, than in the hen of the common species, in which they are represented only by faintly glossed dark spots on most of the feathers.

Practically nothing seems to be known about this species in the wild state, except that it is found in the Peninsula south of Tenasserim, and is suspected of ranging as far north as that district. It is also found in Sumatra. As it was described as long ago as 1760, less than twenty years after the common species, and many skins and a few live specimens have reached Europe, it seems strange that it is still so little known, though of course a forest-haunting bird in a pre-eminently jungly country is not the sort to be easily studied anywhere, and is likely to remain long unfamiliar unless in a district well settled by Europeans, such as much of the range of the grey peacock-pheasant.

Argus Pheasant.

Argusianus argus. *Quou,* Malay.

The true argus pheasant of Malaysia, one of the most remarkable birds known, is the sort of creature which the proverbial blind man in a dark room could hardly miss, for if either party moved about at all he would be pretty certain to come across the enormous centre-tail feathers, twisted-tipped, and between four and five feet long. The wings, however, of the cock argus are still more extraordinary, the secondary quills, which in most birds barely cover the flight feathers in repose and generally expose a good deal of the tips of these, being about twice as long as the "flights" and projecting beyond the end of the body in repose much as the wings of a grasshopper do; in fact, this extension of the wings in the rear and the long straight line they form with the long tail give the bird a remarkably rectilineal outline. and it is anything but graceful, although lightly built and with an easy gait.

The plumage also as seen in repose is nothing striking, being very like a guinea-fowl's scheme of colouring carried out in browns and buffs instead of greys and whites. The hen's is coloured much like the cock's in the exposed portions; neither sex has spurs and both have red legs, and the skin of the head, which is nearly bare, especially in the cock, of a dark blue. The hen's tail is much like a common fowl's, of medium length and folded; otherwise she reminds one more of a small brown hen turkey than of anything else. She weighs about three and a half pounds, and the cock is only about a pound more, although his centre tail feathers bring up his length often to six feet, and the over-developed wing quills make the closed wing nearly a yard long; so, though he is mostly feathers, he looks a very large bird and is nearly as long as any existing species.

The argus is, however, very seldom seen at all in the wild state and has seldom been shot by Europeans. It is never found in open country, and in its chosen haunts, the most dense recesses of the evergreen forests of its home, it has no difficulty in slipping silently away when approached, and it habitually retreats into the thickest cover when alarmed. Even a dog cannot always make them rise, as they are swift of foot; indeed, they probably seldom use their wings except to go up to roost at night, or in case of real need to seek refuge from a quadruped foe. How far the cock could really fly in the open with his peculiar wings, which are constructed all the wrong way for flight—the pinion quills, not the secondaries, being the essential ones—would be an interesting point to solve by experiment, but the exertion of working his great floppy fans would no doubt soon tire him out.

These curious wings, however, form the chief part of the strange and unique display of this bird, which may now and then be seen in captivity, although the argus is not a free shower like the peacock. When displaying, the wings are raised behind and lowered in front, with all quills fully expanded and the first quills touching in front. Now can be seen the characteristic and celebrated ornaments of the male about which so much has been written by theorists on sexual selection: the wonderful rows of eye-spots along the outer webs of the great secondaries, plainly tinted but exquisitely shaded to resemble balls in cups, which everyone must have noticed in these quills when worn in ladies' hats, and the delicate colouring of the pinion-quills with their dark blue shafts, leopard spots, and white-peppered chestnut bands along the inner sides of the shafts. All these artistic excellencies are supposed to please the female, but there is as yet no evidence that she is impressed by them. Her mate is rather handicapped in his manœuvres to give her a good front view by having to keep his head behind the huge screen formed by his wings, which, with the exception of the ends of the long feathers of the raised tail, are all that can be seen from the front; but now and then he pokes his head through from behind the scenes, so it is said. In any case what he presents to the hen is a fan, not a bird, and even in repose, as I have said before, he is not elegant, his decoration being so exaggerated—in fact, he bears the same relation to the peacock that a lady in the silly crinolines once worn does to one gracefully wearing a becoming if lengthy train.

The male argus is a typical crusty old bachelor in his way of living. He selects a small open spot in the forest, and clears off all the dead leaves and twigs and removes all weeds. He is so fussy about keeping this masculine boudoir clean that several of the poaching methods of capturing him depend on this fad of his, the dead-fall or what not being released by his pulling at a peg stuck in his floor, which he immediately tries to remove on coming home. The clearing is no doubt used for a display-place, and it certainly is the male bird's home, for he roosts near it, and is generally to be found in it, when not out searching for food, which consists of insects, slugs, and fallen berries, a fruit very like a prune being an especial favourite. Although these birds do not seem to wander about and fight, they challenge, or at any rate call, quite freely; the note is expressed by the Malay name, but is a two-syllabled one —it always sounded to me like "who-whoop"! The hens are no doubt summoned by this; their own call is, according to Davison, "like how-owoo, how-owoo, the last syllable much prolonged, repeated ten or a dozen times, but getting more and more rapid until it ends in a series of owoo's run together." Both sexes have a bark-like alarm note, and their calls above described can be heard a long way, the male's as much as a mile. Even the two sexes in this bird do not associate constantly, and the hens not only have no regular home, but the breeding-season varies, though eggs are not to be found in the depth of the rains. The eggs are rather like turkeys' eggs, and seven or eight form the clutch. The chicks fledge as rapidly as those of the tragopans, so often miscalled "argus."

The present bird, which ranges to Sumatra, is rather of interest to the naturalist than the sportsman, but the plumage of the cock—at any rate the eyed feathers of the wings—is much used for the plume trade. This use, of course, needs regulating, and the killing of the bird ought to be absolutely prohibited, though it might be permitted to snare it in such a way as not to injure it, and release it after cutting off a two-foot length of the eye-bearing plumes, which could well be spared.

Crested Argus.

Rheinardtius ocellatus.

As a local race of this magnificent bird, previously only known from Tonquin, has of late been found to occur at Pahang in the Malay Peninsula, it may claim brief notice here. It is a bird of general speckled-brown coloration, like the true argus, and the hens are not unlike, except that that of the crested species has a strong bushy crest at the back of the head like her mate; the head is also more feathered in both sexes. The cock has ordinary-sized wings without eye-spots, but a most extravagant tail, several feet long, with broad but tapering feathers marked with brown eye-spots with white centres; a single one of these feathers would easily identify him.

GALLUS FERNGINEUS.

Red Jungle-fowl.

Gallus ferrugineus. *Jungli moorghi*, Hindustani.

"Just like a bantam" is the verdict generally passed on the appearance of honest chanticleer in his wild state, whether the observer be an Anglo-Indian shikari or a lady visiting the London Zoo; and the comparison is apt enough on the whole, for red jungle-fowl, which are simply wild common fowls, have the red-and-black colour in the cock and partridge-brown in the hen, so familiar in many bantams, and are of noticeably small size compared with most tame breeds.

They are over bantam weight, however, cocks averaging about two pounds and hens about half that; and the tail, which is very long in the cock, is carried trailing, not cocked up as in tame fowls. This applies to all kinds of jungle-fowl, none of which strut like the tame bird, and this familiar species has, at any rate in Indian specimens, a particularly slinking gait. Burmese birds have much more the appearance of tame poultry than the Western ones, and are said to be easier to tame; so, unless they are domestic birds run wild, it is probably this particular sub-species that was the ancestor of our farmyard fowls.

To anyone who wants jungle-fowl alive, and wishes to make sure of getting the absolutely real thing, however, I recommend the Indian race, which is characterized by having the ear-lobe (the little skinny flap below the ear) white, and the face flesh-colour, contrasting with the scarlet comb and wattles; the slate-coloured legs are also peculiarly fine. Burmese birds have all the bare skin of the head of the same red, and are certainly not so scared-looking or wild in behaviour, while slightly coarser in form. Of course wherever tame fowls are kept there is a great liability to intermixture with their wild ancestors, so that ill-bred "jungle-fowl" may be expected to turn up anywhere. The fowl also runs wild very readily in the tropics, so that it is really uncertain what its eastern limits are. It does not occur west of India, nor in the south of India itself, neither does it ascend the hills for more than 5,000 feet, and only goes to that level in summer. In the foot-hills it is particularly common, and, generally speaking, it affects hilly country, so long as water is accessible and there is plenty of bamboo or tree-jungle, for it is essentially a woodland bird, though it will come out into the open where there is cultivation in order to feed on the grain. Many of course never see grain all their lives, and live entirely on wild seeds, herbage, insects, &c.

In Burma jungle-fowl are common both in the hills and plains, and extend into Tenasserim and Sumatra. Even if the Burmese and Malayan birds are truly wild, I quite agree with Hume that the genuine aboriginal wildness of the red jungle-fowl found in the East Indies beyond Sumatra is very doubtful. The very distinct green jungle-fowl (*Gallus varius*) ranges from Java to Flores, and looking to the distribution of jungle-fowl and similar birds generally, it is very unlikely that the red species originally lived alongside this bird.

However, to consider more practical matters. This jungle-fowl may be looked for anywhere in the limits above specified if the country is suitable; it avoids alike deserts and high cultivation, and is generally absent from alluvial land, though quite common in the Sundarbans. Here, however, it is suspected of being an introduced bird, as it certainly is on the Cocos.

The fowl since its domestication by man has added no new note to its vocabulary: cackle, cluck and crow were its original language. But whereas the tame cock is always credited with saying "cock-a-doodle-doo," the wild bird's call is better rendered "cock-a-doodle-don't," given in a shrill, aggressive falsetto. Anyone who has heard a bantam crow knows exactly what I mean· for the notes of bantam and wild cock are indistinguishable. Like a bantam-cock, also, the wild bird will live quite happily with a single hen, though this is not universal, and harems are often found; no doubt, as too often with his betters, polygamy is simply a matter of opportunity with chanticleer, though even in the tame state it is often obvious that he has a particular affection for one hen, as was noticed by Chaucer in his "Nonnes Priestes Tale." Jungle-fowl of this species particularly affect sál jungle where it exists, and in India are seldom found away from it; they roost on trees at night, and take to them in any case rather more readily than pheasants. Their flight is also much like that of pheasants, so that they afford very similar shooting if they can be driven; but they will not rise if they can help it, and in thick cover you cannot even see them as a rule without a dog to put them up. They will readily answer an imitation of their crow—at least I found it so the only time I tried; and anyone can practise on a bantam-cock, which will probably attack them when he understands the insult!

Jungle-fowl themselves are exceedingly pugnacious, and have regular fighting-places in the jungle; the duels are sometimes to the death, for the birds have enormous spurs. When challenging, or courting a hen, the wild cock erects his tail like a tame one. After the breeding-season, which may be at any time during the first half of the year, but in the north at any rate only during the second quarter, the cock goes into undress, his flaming frill of hackles giving place to a sober short collar of black, and, as he loses his long tail "sickles" at the same time, he hardly looks like the same bird. Young cocks begin to show the male feathering long before they are full sized, and so are easily distinguishable from their sisters. In the autumn these yearling young birds are fat and particularly good eating.

The jungle-hen lays on the ground in thickets, the nest being a mere scrape among dead leaves as a rule, but some make up a nest of hay and stalks, &c. About half a dozen eggs are the usual clutch, and they are cream-coloured and of course smaller than a tame hen's. The chicks are striped with chocolate and cream on a brown ground; the mother looks after them with the greatest care and devoted courage.

Naturally so widespread a bird as this has many names, mostly signifying the same as the English—wild fowl; *Bon-kokra* in Bengali; *Ayam-utan* in the Malay States; *Tau-kyet* in Burma; *Natsu-pia* among the Bhutias; *Pazok-tchi* with the Lepchas; and *Beer-seem* among the Kols.

Grey Jungle-fowl.

Gallus sonnerati. *Ran-kombadi,* Mahratta.

The grey jungle-fowl, which takes in the south of India the place occupied by the red species in the north, is so distinct from this that a single feather would in many cases identify it. The cock is grey, the feathers both above and below being narrow, pointed, and with white shafts. The neck-feathers have bright yellow tips, and there is a patch of orange on the wing, these yellow or orange tips being solid, not split up into barbs like the rest of the feathers. In some individuals the tips of the neck-feathers are white instead of yellow.

Those interested in fly-fishing probably already know this jungle-cock's hackle by sight, as it is one of the standard feathers for fly-dressing. After breeding they are replaced for a time by an undress collar of sooty black, and in young cocks this black neck is the first sign of masculine plumage to appear, so that in a flock they may easily be known from hens, whose necks are yellowish, though not so bright as the distinct yellow and black seen in the neck of the red jungle-hen.

In her upper plumage generally, however, the grey jungle-hen is much like the red, being of a similar brown, but underneath she is very different—pure white, regularly edged on each feather with black. To those who know tame poultry she may well be called to mind as a bird with a Brown Leghorn hen's plumage above, and a Silver Wyandotte's below. Cocks have red legs, and hens and young birds yellow.

In weight this species averages a few ounces more than the last, and it is more strongly built, but it seems to be far more timid and less plucky in disposition, although now and then found fighting furiously. So wary is it, and such a runner, that it affords but little sport unless it can be driven when working the smaller sholas in the Nilgiris, where it is a well-known bird—in fact, it is common all through the hill ranges of Southern India, and ranges occasionally as high as seven thousand feet. It likes thin rather than thick jungle, and is especially attracted when bamboo or the strobilanthes under-

1/5 GALLUS SONNERATI

growth is seeding. In the rains it spends a good deal of time in the trees even by day, and always roosts there.

It is not nearly so sociable as the red jungle-fowl; a party will only consist of an old pair and their young, and old cocks, which are particularly wary, are often found alone.

The note is just as characteristic in this species as the plumage; the crow is difficult to recognize as such at first, until one observes the deliberateness and periodicity of its production. I can only describe and imitate it by putting in words how it struck me when I at last caught a captive specimen in the act of challenging: "Oh lor'! what a *cac*-kle!" Once known, however, it can be recognized at a long distance; but the birds only crow when in full feather, *i.e.*, from October to May.

The cackle of the cock is more easily compared to that of the red jungle-fowl; it seems to correspond to the last two notes of the familiar "tuk-tuk-tuk-tuk-tuk-auk" of agitated poultry, and sounds like "tooruk, tooruk," pronounced in a harsh tone. The hen is rarely heard, but is said to be voluble when she does call, uttering a hoarse note like "uk-a-uk-a-uk" very rapidly, no doubt very like the cock's alarm-calls.

On the western face of the Nilgiris these birds breed during the last quarter of the year, but the time is different in different localities, and somewhere or other they may be found nesting in almost any month. The eggs incline to be more numerous than those of the northern jungle-fowl, and are far more variable, being either short, with a coarse pitted glossy shell of rich buff, or long, fine-shelled and pale creamy, or some type between these, often brown-speckled.

They have bred in captivity in England pretty freely, and in one case were crossed freely with game bantams, the resulting hybrids breeding freely every way. The same experimenter who bred these also bred the pure birds and turned them out in the woods, but found them too quickly destroyed by foxes to render the process worth keeping up; but he mentioned that they got up wilder than the pheasants, and afforded better sport, being quicker and more difficult to hit.

Davison, who was much impressed by what he believed to be the peaceful disposition of the species, says that they would not breed in captivity (in India), and cocks and hens would live together in peace; no doubt this was simply because his birds never got into breeding form. Personally, from what I have seen of captive birds, I should never dream of letting two cocks be together along with hens. The ranges of the two species of Indian jungle-fowl meet in places, and in a bit of sál jungle near Panchmarhi the red bird is found in the middle of territory occupied by the present kind. Jerdon once shot a hybrid at one of the meeting points—on the Godavari where it joins the Indravati—so that it is as well to mention, for the better recognition of such, that the peculiar horny spangles of the grey bird's hackle are lost in the cross, but that the characteristic light shafts to the plumage of the cock and light centres to the hen's breast feathers show up very distinctly; at least that was the case with a pair of hybrids bred between a Sonnerat or grey cock and a mongrel bantam hen in the London Zoo in 1913. The cock hybrid bird's crow was also distinct, in four syllables, "cock-a-doo-doo." In general appearance he was very like a mongrel reddish bantam, but had the breast reddish as well as the back, and no grey anywhere, this being, so to speak, overlaid with orange-red.

Ceylon Jungle-fowl.

*Gallus lafayettii.** *Weli kukula,* Cingalese.

Even if it were not confined to Ceylon, and the only species of jungle-fowl found in that island, there would be no difficulty in distinguishing the Ceylon jungle-fowl.

The cock's plumage, red below as well as above, and with the same narrow feathers, glassy-lustred everywhere, is quite distinct from that of either of the mainland species, to say nothing of the yellow patch in the middle of the red of his comb.

The hen, like the cock, shows her distinction from the red jungle-fowl of the north in her under plumage chiefly; this, instead of the fawn-colour found in the hen of the red jungle-fowl, is black-and-white, not in the form of white centres and

* *stanleyi* on plate.

1/4

GALLUS STANLEYI.

black edges to the feathers as in the hen grey jungle-fowl, but irregularly mottled and intermixed with brown. Her comb is particularly small even for a wild hen's, and her face feathered like a partridge's, not bare as in the hen of the mainland jungle-fowl.

Young cocks can be distinguished from hens by being more reddish on the brown upper parts and having only black and brown below, with no white.

The voice of this jungle-fowl is quite as distinct from that of the two mainland birds as his plumage is, if the words "George Joyce" or "John Joyce," the renderings given of it, are at all correct. A bird in the London Zoo, believed to be a hybrid Ceylon common fowl, crowed in three syllables "cock-a-doo."

Hybrids between the jungle-fowl and tame poultry are liable to occur, as the wild bird sometimes crosses with village hens, being able to overcome their consorts; so the characteristic points of yellow-patched comb and glazed lower plumage should be borne in mind in determining the characteristics of a genuine bird.

Many tame cocks of a red colour—if not most, in places where poultry breed anyhow—have reddish-brown instead of black breasts, but on examination it will be seen that the feathering here is ordinary, not glazed like that of the upper parts, so that there is no reason to believe that such birds have a cross of the Ceylon wild fowl. Similarly, the grey domestic fowls, which are also common, are never marked in detail like the grey Sonnerat cock, nor have they his peculiar pointed feathering.

Hybrids with the Ceylon jungle-fowl and common fowl, by the way, have been proved fertile, with one of the parents at least.

In Ceylon this bird is very generally distributed in all jungly portions, but favours low rather than high ground in the north, and in the south is scarcer and more a bird of the hills. It likes a dry soil and scrub-jungle, especially of thorn and bamboo. The cocks are far more often seen than the hens, though no doubt the inconspicuousness of the latter has a good deal to do with this; but in any case they are shyer in disposition. More than one

12

hen and brood often associate, and flocks of these jungle-fowl may sometimes be seen feeding on cultivated land, but on the whole the species, like the grey jungle-fowl, seems to be more shy and unsociable than the red bird of the north. It also resembles the grey jungle-cock in taking a good deal to trees in wet weather, and in its fondness for *Strobilanthes* seed, the "nilloo" of Ceylon, on which the bird feeds so greedily that it seems to become stupefied, being a plant of this genus.

There is a point of affinity with the red jungle-fowl, however, not only in the colour of the male of this bird, but in his habit of flapping his wings before he crows; this, as far as I have seen, the grey bird does not do, but in this species, as in the red jungle-cock and his tame descendants, the habit must be very pronounced, for shooters can and do decoy the cocks within shot by imitating this sound, the native doing it by striking the thigh with the slightly-curved open hand. Any other noise will cause the wary bird to run off at once.

The food of this bird consists chiefly of wild seeds, but also of insects, especially of white ants. In one part of Ceylon or the other it may be found breeding at any time of the year, depending apparently on the incidence of the north-east monsoon; it is even thought that the birds may be double-brooded, and they seem to pair. The young, even the full grown, have been seen to show great reluctance to leave their dead mother when she had been shot.

The small clutch of two to four eggs is laid on the ground in a scanty nest in some thicket or at times on a decayed log, are much pale buff, and speckled finely and spotted with rusty red, like many eggs of the grey jungle-fowl, to which this species, in spite of its red colour, is probably quite as nearly related as it is to the red bird, unless it represents the ancestor of both, as seems possible from the hybrid grey and domestic specimen before-mentioned, having come out reddish below as well as above. Moreover, this hybrid, like the grey jungle-fowl, had a purple, not green, tail, and this is also the case with the Ceylon jungle-fowl.

The legs of this species are also said to be yellow, but, Lewis Wright, in Cassell's "Book of Poultry," says, pink in the cock, which is far more likely to be correct, for the cock grey jungle-fowl has salmon-coloured legs.

The female of this species is called in Cingalese *Weli kikili*, and the Tamil name is *Kaida koli*.

Red Spur-fowl.

Galloperdix spadicea. *Chota jungli murghi*, Hindustani.

The scientific name "cock-partridge" and the Hindustani one, "little jungle-fowl," give a very good idea of the character of this queer little wild bantam, though it is a bantam *hen*, not a cock, which it resembles, the tail being short and hen-like, while there are no hackle-feathers. The comb is also wanting, but the eyes are surrounded with a red bare skin, and the feet and bill are also red. The hen is also not so very unlike some fowls, a light, sometimes greyish brown, more or less pencilled across with black, but the cock is of a strikingly distinct colour, being of an almost uniform chestnut throughout, though this again is much like the shade of the much-boomed "Rhode Island Red" poultry.

Although lacking the distinctive decorations of their aristocratic relations, the jungle-fowl, spur-fowl easily surpass them in the practical matter of armature; the cock has usually two spurs on each leg, sometimes more, while it is a poor hen that cannot raise at least one spur on one leg, and some have two on one and one on the other.

The distribution of this bird is curious; it is scattered about here and there throughout the Indian Peninsula; yet though it does not extend north of the Ganges in this region, it turns up again in the Oudh Tarai. It is essentially a bird of hilly and rocky jungle, and is never found in flat country or open land of any sort. It is very shy, seldom coming into cultivation, and even when its haunts are invaded always greatly prefers running to flying. It is very swift on foot, and even a dog has difficulty in putting it up; when it does rise, it goes off with a whirr and loud cackle, and is easily shot, but not at all easily retrieved if not killed, as it goes to ground like a rabbit. In fact the best way of getting it is to treat it like one, and shoot it running, as Hume says. It is a perching bird, roosting at night, and being fond of taking to a thick bush when put up by a dog, a refuge from which it is most difficult to dislodge it. In compensation for its extreme aversion to giving a sporting chance to the gunner, it is an excellent bird for the table, and Hume considers it best of the Indian partridge tribe. But it is hard to get many of them ; about two or three in a day's shooting is about what may be expected on the Nilgiris, where they range up to 5,000 feet, or even over. Although often in coveys of four or five, they even then do not go off all at once, but now and then, and here and there, and they are frequently found in pairs, sometimes even alone.

The cackling call which re-unites a scattered flock is said to be much like that of a hen, and they are credited with a crowing call, which seems, however, to be rarely heard. Their chief food is jungle berries, seeds, and insects, but they will occasionally come into fields for grain, and they seem to need water frequently, as they are constantly to be found near it, and a thorny ravine with a stream in it is the surest locality for them.

This bird is suspected of breeding twice in the year, but the only certain season is during the first six months ; the nest is on the ground, in cover, and the eggs are rather like small hens' eggs of the brown-tinted variety so much esteemed; they vary a good deal in shade, and seven or eight seems to be the maximum set, though smaller and larger numbers occur.

The red spur-fowl is fairly well off for names ; if its Mahratta title, *kokatri*, does describe its call well, it certainly must have a note that can be fairly called a crow. The Deccan Mahrattas, however, call it *Kustoor;* in Telugu it is *Yerra* or *Jitta-kodi*, and in Tamil *Sarrava koli*.

Painted Spur-fowl.

Galloperdix lunulatus. *Askal*, Orissa.

The white-spotted cock of the painted spur-fowl is a very pretty and distinct-looking bird, as far as the plumage goes, but the absence of the bright red colouring of the bill and feet, which are dull blackish, and the very faint indication of red round the eye, make the dull, plain, brown hen a very ordinary looking creature; her only noticeable point of colour is the chestnut face.

The ground-colour of the cock's plumage is chestnut above for the most part, but the crown and shoulders are glossy green-black; on the buff breast the spots are black instead of white, and the head and neck are all black and white.

The Askal is generally seen when rocky hills are being beaten for big game, such places being its usual resorts, but they must have plenty of vegetable cover as well as stones. Even when thus forced out, the birds will only fly once, going to ground among the boulders when they have had one flight; on the wing they look much like jungle-fowls and, unlike the last species, do not drop readily to a shot; they resemble it, however, in their speed of foot.

This also is a very local bird, and although its area of habitat is much the same as that of the red spur-fowl, it is not found in the north-west, while it extends east to Bengal, and is wanting on the coast of Malabar. Moreover, in localities where the red spur-fowl is found this species is generally wanting, and *vice versâ*, so that on the whole their distribution does not coincide nearly so much as might be imagined from the consideration of their range as a whole. The choice of station, too, is somewhat different in the two species, since, though both love hills and thick cover, the present one is more distinctively and exclusively a rock-haunter. The call is said to be a peculiar loud *chur*, *chur*, *chur*, anything but fowl-like, but Jerdon, in speaking of the "fine cackling sort of call, very fowl-like," attributes this to the males, so that no doubt the challenge and alarm-notes are, as one would expect, quite different. Not only is the bird very shy and hard to get, but it is not so succulent and gamey in character of flesh as the red spur-fowl. The cocks have been found in confinement to be very pugnacious, and their black legs are as well provided with spurs as are the red ones of the other species; hens also are usually armed in the same way, with a pair of spurs or a single one.

The five eggs, which may be found as early as March or as

late as June, are laid on the ground; they are buff in colour. In spite of their natural shyness, the old birds show great boldness when accompanying chicks; they will try to draw away pursuit by artifice, and if a chick be captured will come within a few yards.

This species is well distinguished by names; though in Telugu it shares the name *Jitta-kodi* with the last, it is called *Kul koli* in Tamil, while the Gond name is *Hutka* and the Uriya *Kainjer*.

Ceylon Spur-fowl.

Galloperdix bicalcarata. *Haban-kukula*, Cingalese.

It is not only in appearance that the spur-fowls are like jungle-fowl, but in distribution, there being two in India and one in Ceylon, while there is a curious coincidence in colour, the most northern and southern species of the jungle-fowl having red plumage, while in the spur-fowl it is the red species, which alone is found in Northern India, and the present southernmost one, which have the red legs and conspicuous red skin round the eyes.

As to plumage the cock of this bird is not unlike a richly coloured edition of the hen of the grey jungle-fowl, having the neck and flanks white with black edgings to the feathers, these edgings failing on the breast; the back shows the characteristic spur-fowl chestnut. The brown plumage of the hen is rendered distinctive by the contrast of the red eye patch and extremities, otherwise she is more like the hen of the painted than of the red species—another coincidence with the jungle-fowl, in which it will be remembered the hen of the Ceylon species is more like the grey than the red jungle-fowl hen. The white spots on the cock's wings also recall the markings of the painted spur-fowl.

The Ceylon spur-fowl makes itself well known wherever found by its cackling call, the notes being on an ascending scale, and beginning at about 6 o'clock in the morning; but as it is a decided ventriloquist, very swift-footed, and an adept at taking cover, hearing it is one thing and getting a shot at it another.

It especially affects hills, but must always have cover, and hence its absence from the north of the island is not surprising, as the jungle there is too open for its tastes. In many places it is quite abundant, and is always found in small coveys, which, as Legge suggests, are no doubt families. The cock calls a scattered covey together by a pipe like a turkey-chick's, which changes to a louder whistle as the birds get an answer or become more confident. The cocks are very quarrelsome with each other, and, as in the other species, have several spurs, while the hens are also spurred as a rule.

Layard considered this bird very good eating, and much resembling grouse; it weighs about twelve ounces if a cock, but the hens are about three ounces lighter. The four creamy eggs have been found in February, May, July, and October; they are laid on the ground in forest.

Common or White-crested Kalij.

**Gennæus albocristatus.* *Kalij*, Hindustani.

Of the narrow-crested, fowl-tailed, red-faced, black and white hill pheasants known as kalijes, the present is the best known, occupying as it does, a great range from the borders of Afghanistan to those of Nepal, and being seen near the ways and works of men more than any other pheasant. From all his relatives and from all other Indian game birds, the cock's long but skimpy white crest will distinguish him; the blue-black of his upper parts gets very rusty about the shoulders, and is diversified by white bars lower down the back; underneath he is of a very soiled white and whity-brown, the feathers here being long and pointed, as in all these white-breasted kalijes. His whitish legs are spurred.

The hen is of a type very distinctive of this group. She is crested and red-faced like the cock, and has the tail but little shorter. Her brown plumage, though with pale edges, is only really diversified by the black outer tail feathers.

---

* *Euplocamus* on plate.

In its Himalayan home this kalij chiefly frequents the middle and outer ranges, and is also found in the Siwaliks, alone of all the Himalayan pheasants. In winter time it comes down near the roads and cultivation, ranging in summer up to the haunts of the moonal and tragopan, even as high as 10,000 feet; but, generally speaking, it may be said to inhabit an intermediate zone between these and the jungle-fowl and peafowl of the foot-hills.

It always keeps near or in cover of some sort, but prefers low to high jungle, and especially haunts wooded hollows and ravines, even in the interior, where it may be found in any sort of forest; it does not as a rule go into woods far from human habitations, even the former traces of man's occupation being an attraction to it. Yet, like that most domesticated of birds, the house sparrow, it does not bear confinement at all well; such birds probably know or suspect man too much to be happy when in his power.

Its desire for grain, which it can generally procure in human neighbourhood, especially from the droppings of domestic animals, is no doubt the great reason for this approach to the enemy of its kind, but it also, of course, feeds largely on the shoots, berries and insects that form, as it were, the standard natural food of pheasants.

Although three, four, or a dozen may be found near each other, the birds are not really gregarious, and when breeding go in pairs; moreover, the cocks are exceedingly pugnacious to each other. Their challenge, common to all the group, is a peculiar drumming made by rapidly whirring the wings. The call is a sharp *tweet-tweet* or whistling chuckle, given out on rising, and continued excitedly when the bird is treed by some terrestrial foe. When thus treed the kalij is far from being brought to book, for it often keeps a wary eye open, and when discovered will drop down on the wrong side of the tree for the gunner and make off. Its flight is exceedingly fast, but it travels fast on foot also, and unless it has not been worried by man, and so is fairly steady when treed, is not easy to get in any number, and so falls under the head of casual game rather than a regular sporting bird. I can find no note on its edible

4
EUPLOCAMUS LEUCOMELANOS.

qualities, but the group generally are not better eating than ordinary tame fowls.

These birds breed from the Tarai to an elevation of 8,000 feet, so that it is not surprising that the eggs may be found, according to elevation, from early in April to late in June. The nest is well hidden in low cover, such as grass or fern, but is very slight as a rule. The sitting is usually nine, and the colour is some shade of buff. They are about the size of small hen's eggs. The hen sits, says Hume, for rather over three weeks, and the cock keeps with her and the brood till they are nearly full grown. The mature weight of this bird, by the way, is rather over two pounds in the cock and about half a pound less in the hen. In the North-west Himalayas the sexes are discriminated by name—*Kalesur*, applying to the cock, and *Kalesi* to the hen, while *Kolsa* is the Punjabi and Chamba name for the species. As wild hybrids are very rare in India, it is worth mentioning that Hume once shot a male bird which he thought must have been a cross between this and a koklass, and Captain Fisher got one which had the head, neck, and crest of the kalij, while the back and alternate feathers of the tail were like a monal's.

Nepal Kalij.

**Gennæus leucomelanus.* *Rechabo*, Bhutia.

It is not at all surprising that this species in Hindustani simply shares the name of *Kalij* with the common white-crested bird, for except that the present bird has a black crest, not quite so long and therefore not drooping, the two are practically alike, blue-black above and dirty-white below, with the rump transversely barred with pure white. The kalij of Nepal, however, which is not found elsewhere, and is at any rate, except, perhaps, on the extreme eastern and western ends of that kingdom, the only kind found there, is not quite so rusty-looking above as the white-crested, nor so stout and pale in the leg, nor is it quite so large in most cases.

**Euplocamus* on plate.

The hen is a brown, narrow-crested, hen-tailed bird like the white-crested kalij hen, but it is slightly darker, with a shorter crest, which does not show the greyish tinge found in the crest of the hen of the other species; but the hen kalijes of this type can hardly be separated with any readiness or certainty, at any rate by a beginner. The young cock, gets his full plumage during his first year, when about five months old; three months' old chicks are brown with some black bars above.

The Nepal kalij is much the commonest of the pheasants of its native state; it is strictly a hill bird, with a rather limited vertical range, never going down to the Tarai region, and rarely ascending over 9,000 feet. It keeps to thick forest and is a great percher, not only roosting on trees, but being commonly met with perched in them as one makes one's way through forest, according to Dr. Scully. It may be, however, that the birds seen so much aloft have simply "treed" through alarm in many cases. When approached, Dr. Scully says, they fly rapidly down and run off. He found the best plan to shoot them was to wait, in winter, when the birds come down to the foot of the hills near the trees to which they resort to roost, though occasionally a shot could be got at one as it crossed a path. He found the birds stood captivity well, and he reared chicks to maturity, which conflicts with what others say about the difficulty of keeping the pheasants of this group. But in these matters a great deal must be allowed for skill, and Dr. Scully, as a medical man, would naturally bring more intelligence to his task than the ordinary "man in the street" who is generally a hopeless bungler with live stock, even if he is interested in them from a sporting or natural history point of view, unless he has had some experience with tame things.

The birds usually go in pairs or small parties up to ten in number. All that is known about the breeding is that a chick so young as to measure only two inches in the closed wing, is recorded by Scully as captured in June; it was rufous brown on the head and dirty buff below, with no stripes apparently.

A W Strutt del.

F Waller Chromo lith. 18 Hatton Garden London.

EUPLOCAMUS ♂ MELANOTIS

Black-backed Kalij.

**Gennæus melanonotus.* *Muthura*, Bengali.

The black-backed kalij of Sikkim, although generally similar to the last two species, blue-black above and dirty white below, differs from both in having no white at all on the upper surface, not only the crest, but also the rump being blue-black continuous with the rest of the upper parts, whose silky purple seems to me particularly uniform and rich in this species. The crest, as in the last species, is not so long as in the white-crested kind; but the bird seems to be quite as large. The weights of these kalijes, however, intergrade so that size cannot be considered of any importance in dealing with them.

The hen is of the same sober brown as that of the other two, with narrow crest; but like that of the Nepal kalij, she is rather darker altogether than the hen of the white-crested bird. The legs are described as pale horny brown, darker than those of the white-crested.

This kalij extends into Bhutan on one side of its range, while on the other it encroaches on Nepal, but its characteristic home is Sikkim; its Lepcha name is *Kar-rhyak*, *Kirrik* in Bhutan. It ranges from quite the foot of the hills up to 6,000 feet, and is common in tea gardens, or used to be, but more than a generation ago Hume noted that the garden coolies used to find its nests among the tea and destroy its eggs, so that he anticipated it would become comparatively rare, especially as it was inclined to affect the outer hills which were being taken up for tea-cultivation rather than the interior.

It always keeps to cover of some sort, and is just as much at home among the tea as in jungly growth; ravines well bushed over are its favourite haunts. Gammie found it very tame in Sikkim, so that when met with feeding on the roads in morning and evening it would only walk out of the way when disturbed. During the day it shuns the sun, and seldom perches unless put up by some enemy on the ground. Human intruders it avoids

---

* *Euplocamus* on plate.

by running, or if that is impossible takes a short low flight and settles on the ground again. Its alarm note is given by Gammie as *koorchi, koorchi, koorchi,* while the challenge call is *koor, koor* and the fighting note *waak, waak.* The same drumming with the wings as is indulged in by other kalijes is also performed by this one, and the natives, apparently with reason, regard it as a presage of rain.

It is a very omnivorous bird, eating all sorts of insects, except ants, which the natives told Gammie were refused by captive birds; beetle grubs and wild yams, and the fruits of the totney and yellow raspberry, are favourite articles of food, and grain of all sorts is readily devoured, with the shoots of nettles and even ferns. The flesh is not very good, and the bird affords little sport, being a great runner, and affecting cover so thick that even a dog can do little in it.

About Darjeeling it has been noticed to be very constant to its roosting-trees and even keeps to the same bough, so that it is easily located by its accumulated droppings. It generally goes in pairs or only three or four together, and the cocks fight much in the breeding season.

Although in the higher parts of its range hard-set eggs have been found at the end of July, at the other end of its zone, low down, they may be laid in March, no one seems to have seen any sort of a nest constructed, the eggs being laid in grass or under cover of bush, fern, or rocks on the ground itself. Hume never heard of more than ten eggs in a clutch, and their colour varies from pale brown to pinkish cream-colour.

Purple or Horsfield's Kalij.

**Gennæus horsfieldi.* *Dorik,* Assam.

Like the last species, this bird has the native name *Muthura,* and it is certainly allied to it, though more nearly to the next to be mentioned. It differs from the three most typical kalijes in having the underparts black with feathers of the ordinary rounded

* *Euplocamus* on plate.

EUPLOCAMUS $\frac{1}{3}$ HORSFIELDI

shape, not pointed; from this it is often called the black-breasted kalij, a rather misleading name, as it gives the impression that the black breast contrasts with the upper surface, which is not the case. In fact, the bird is the most simply and uniformly-coloured of all our pheasants, its purple-glossed black plumage being only relieved by white bars on the lower back. The crest is long and narrow.

The hen bird is very similar to the hens of the three white-breasted kalijes, both in plumage and crest; the only special point she shows is the contrast between the reddish-brown of the central tail-feathers with the more olive-brown of the rump; but the difference is slight, and she can hardly be picked out from the others above-mentioned, while curiously enough she has no such near resemblance in colour to the hen of the lineated pheasant next to be dealt with, a much closer ally.

The purple or black-breasted kalij is a hill-bird like the group generally, and ranges from Chittagong to the Daphla hills and Eastern Bhutan, extending also to the northern parts of Arrakan and to Burma as far as Bhamo; Southern Manipur is also within its range, but its exact limits are not very easy to determine, as interbreeding between it and other forms undoubtedly goes on. It is not so high in its range on the hills as the white-breasted kalijes, seldom going above 4,000 feet and haunting jungle at the edge of cultivation and along rivers. Except for this, its habits, like its size, show no particular distinction from those of the three previous kalijes. It keeps mostly to cover, and only shows sport when hunted up with dogs, when it often takes to trees. Well-wooded hills and ravines are favourite resorts, and only a few birds are seen together, pairs being more usual, though as many as eight or even eleven birds have been seen in a party. When flushed it rises noisily, and with a shrill repeated *cheep*.

The cocks are exceedingly pugnacious; two have been found fighting with such fury that both were captured by hand, in a much pecked and exhausted state; and one has been seen to stand up for some time to a red jungle-cock, which won in the end, the *casus belli* having been a white ant heap, on the swarming inmates of which these cantankerous birds could not agree to dine in peace. The flesh of this kalij, by the way, although white, is not as good as that of the jungle-fowl; a cock may weigh about three pounds, but is usually less. Besides insects, they feed on worms, shoots, and grain, for which they will often scratch in horse-dung. They are reported very difficult to tame, but nevertheless the species has often been brought to Europe, and has been during the time of writing represented at the London Zoo. I notice that when frightened the bird raises and spreads out his thin stiff crest horizontally, so that it is very broad and conspicuous.

The eggs of this bird may be found from March to June; they are warm light brown to pale buff in tint, and strong shelled; four have been found in a nest, which is made of dry leaves on the ground.

In the Garo Hills this species is known as *Dúrug* or *Dirrik*.

Lineated Kalij.

*Gennæus lineatus. *Yit*, Burmese.

The lineated kalij, which closely resembles the purple kalij in size, form, and even the habit of erecting and spreading out the crest, is the most westerly of the group of black-breasted, pencilled-backed pheasants which culminate in the well-known silver pheasant of our aviaries, which is a South Chinese bird. It is generally distributed over hilly country in Burma, and is sometimes called the Burmese silver pheasant, or even, as by Hume, by the very awkward name of "vermicellated" pheasant. "Grey-backed kalij" would really be the best name for it, as it is a typical kalij in size and shape, hen-tailed and narrow-crested, while the most striking point about it is the contrast between its delicate grey upper surface and black crest and under-parts. The pure white along the upper half of the centre tail-feathers is also a striking colour-point, and with the scarlet face, goes to make up a singularly handsome, if quiet-looking bird.

For comparison with other races it should be noted that the grey of the back is not a solid colour, but made up of very fine

* *Euplocamus* on plate.

EUPLOCAMUS LINEATUS

pencilling of black and white lines, such as is seen on the backs of many of the males of the duck tribe, but very rarely elsewhere; it is irregular and does not follow the edges of the feathers.

The hen bird is quite like the hens of the white-breasted and purple kalijes in form, and is also brown above, but her under plumage and neck are different, as are also the outer tail-feathers, being variegated, the former with well-marked white streaks, the latter with tranverse pencillings of white on the black ground.

The lineated kalij, like the purple, does not range high up, even 4,000 feet being generally higher than it cares to go, while it has no objection to sea-level if it can get suitable jungly cover and ravines or similar declivities. What it especially likes is long grass, bamboos, small trees, and brushwood, on hill-sides; and it prefers deciduous-leaved trees to evergreen forest. On account of the steepness and treacherous character of much of the ground it frequents, it is often not easy to shoot, and is a great runner, though a dog will put it up readily enough.

It has, in fact, the regular kalij habits; it is, for instance, usually found in pairs, though broods may keep together. The cock challenges by whirring with his wings; the alarm-note is a whistled *yit*, whence no doubt the native name in Burmese. In Arrakanese the name is *Rak*, in Karen *Phugyk*, while the Talain name is *Synklouk*.

It is a mixed feeder, but has a special liking for ants, black as well as white, and for the figs of the peepul; in places where it can get the succulent shoot of a certain orchid to feed upon it can do without water for some time, but usually likes to be near it, drinking at about 10 a.m. In some localities it avoids cultivation altogether, in others it will come freely out into rice fields to obtain grain. It also feeds on young leaves and grass.

One curious habit, observed by Colonel Bingham, is that it often comes into clearings on bright moonlight nights, a most curious trait in a pheasant or any nearly allied bird. Chicks are said to be hard to rear, but the species has often been brought to Europe, and, like the purple, was on view at the London Zoo at the time of writing.

It has hybridized in captivity with the Chinese silver pheasant, the resulting hybrid being practically identical with the doubtful form known as Anderson's silver pheasant (*Gennæus andersoni*). Owing also to hybridism in the wild state, both with the purple kalij and silver pheasant, the limits of this bird are hard to fix. Eggs of the lineated kalij may be found from March to May, in a hollow scratched out among dry leaves or scratched in the ground and lined with such leaves, but are generally well hidden. The eggs are seldom more than eight and are of a buff or stone colour with a pinkish tinge.

Silver Pheasant.

Gennæus nycthemerus.

The lovely silver pheasant, for the last century domesticated at home, is not known as a wild bird in India, but as he lives as near as South China, and his hybrid offspring infest our bordering states, to the great bewilderment of sportsmen and naturalists, he comes into the tangled tale of the pencilled-backed kalijes. The cock silver pheasant has a folded tail, but it is long, and the top feathers are so long and arched that the general effect looks quite different from that of the tail of other kalijes. These top feathers are pure white, and white is the ground colour of the rest of the plumage except for the blue-black underparts and crest. The black pencilling is regular, but extremely fine and inconspicuous except on the wings and side tail-feathers. The crest droops as in the white-crested or common kalij, but is far fuller than in that or any other species. The legs are red as well as the face in both sexes. The hen has a short crest, hardly noticeable, and black; and perfectly plain brown plumage, the only markings being irregular black and white pencilling on the outside tail feathers. Young birds have this pencilling on the breast, and the cock, which does not get his full colour till the second year, goes through a most peculiar series of changes before and while moulting into it, the feathers appearing to change colour without a moult to some extent. A maturing specimen of this sex might easily be referred to half a dozen species at various times, as species have been reckoned in this group.

The form known as Anderson's silver pheasant (*Gennæus*

EUPLOCAMUS ANDERSONI

andersoni) from the Kachin Hills and the Ruby Mines, seems simply to be a hybrid between this bird and the lineated kalij; but the bird figured by Hume as Crawford's silver pheasant (*Euplocamus andersoni* on plate) seems to me the same bird with a further cross of the lineated. As further intermediate forms occur between these half-silver pheasants and the purple kalij, and as that bird also undoubtedly grades into the lineated kalij where their ranges approach, through more interbreeding, it will be seen that it is very difficult to draw any lines between all these black-breasted kalijes, silvered or plain; and we may ultimately have to come to the astonishing conclusion that they are all of the same species, of which the type will have to be the Chinese silver pheasant, as the oldest kind known.

Fire-back Pheasant.

**Lophura rufa.* *Mooah-mooah*, Malay.

The splendid fire-back pheasant, distinguished from all other members of the kalij group, to which it belongs, by its sky-blue face, high bushy crest, and large size, which even in the hen goes up to three and a half pounds and in the cock may reach five, is only found in Southern Tenasserim, being one of the many Malay forms which just penetrate our dominions in that direction. Its eastern limit is Sumatra, an allied but distinct species replacing it in Borneo.

Living where it does, this species is not likely to be confused with the Himalayan monal, also a big pheasant with a blue face, but utterly different in every other way. The cock fire-back, in addition to the splendid patch of colour which gives him his name, is remarkable for the lustrous navy-blue of his plumage, set off by white flank-splashes and the white centre-feathers of the curved but folded and hen-like tail, and the coral-red legs, armed with great white spurs. The hen is quite as distinct in her way, on account of her bright foxy-red plumage, marked with black and white below; her legs are red like the cock's, whereas those of the Bornean fire-back are white in both sexes.

* *Euplocamus vieilloti* on plate.

Her crest is quite well developed, though not so large as the cock's.

In Tenasserim, Davidson found this bird associating in small parties, consisting of a male with his harem, though solitary males sometimes occurred ; they always kept to the cover of the evergreen forests, and scratched a good deal. Their food was the usual mixed diet of pheasants—leaves, berries, and insects. When alarmed, the covey ran off together, but could be put up by a dog, when they would fly strongly for a couple of hundred yards and then settle and begin to run again.

The cocks frequently challenged in the usual manner of the group, by whirring with their wings ; and that they are as wantonly vicious in their wild state as they are in captivity was proved by Davison having seen one repeatedly drive a cock argus from his bachelor sanctum ; the poor bird, though he would come back at the bully's whirring challenge, being naturally afraid to stand up to his formidably armed and active antagonist.

Besides the wing-buzzing, the cocks have a vocal alarm-note, which Davison compares to that of the big black-backed squirrel (*Sciurus bicolor*), that fine fellow as big as a cat which is so conspicuous in the forests ; the hens also have the same sharp cry. The egg is known to be buff, and very like some hen's eggs, as these birds have laid in captivity ; but no eggs seem to have been taken in the wild state, although the breeding-season appears to be known, and is said to be in the monsoon.

White Eared-Pheasant.

Crossoptilum tibetanum.

This beautiful large white pheasant, with its snowy loose plumage so well set off by its purple-glossed tail and red face and legs, has never been actually taken in British territory, though it is suspected of occurring on some of the Bhutan passes. It is really a Tibetan and Chinese bird, and Hume was induced to figure it chiefly in order to reproduce a copy of the figure of it given by Hodgson, who described it in 1838, and to quote his description, which was buried in one of the earlier volumes of the Asiatic Society's Journal.

1/6

EUPLOCAMUS VIELLOTTI

There is no need to repeat this description, as the bird is so distinct that if it does turn up in British territory it cannot be mistaken for anything else; but in it Hodgson seems to have been curiously mistaken in two points. He says, first, nothing at all about the curious erect white ear-tufts which the bird has like the few other members of the *Crossoptilum* group—though it must be admitted that they are shortest in this species—nor does his drawing show them. He also describes the bird's tail as " broadly convex, without any sign of the galline compression and curve," the fact being that in all these eared pheasants the tail *is* folded and fowl-like, with the top feathers curved, and even looser-webbed than the rest. Hodgson's bird was got from a Nepalese envoy who had been to Pekin, but it is now too late to ask where *he* got it!

In China, according to Père David, this bird inhabits bushy localities and is very sedentary and sociable, even during the breeding-season. Being poor eating, and respected by native superstition, it has a better chance of survival than its near relative, the only one usually seen in Europe, the brown crossoptilum (*C. mantchuricum*) which he regards as in danger of extinction, owing to persecution and the cutting down of the forests. The white crossoptilum has been exhibited at the London Zoo, but I never saw but the one specimen.

Koklass Pheasant.

Pucrasia macrolopha. *Koklas*, Hindustani.

Sometimes called the Pukras, from another native name *Pokras*, this pheasant is very distinct in type from all our other species; the tail, though short for a pheasant, is pointed, and the head provided with a long crest, in three portions, for only the central part grows from the crown, two longer tufts proceeding from the sides of the head, which is deep green in colour except for the central crest, which is pale brown. Just where the head joins the neck there is a long oblique white spot on each side.

The colour of the body-plumage, which is long and pointed,

and longitudinally streaked for the most part, varies much according to locality, and several species used to be distinguished on this account, though so much variability occurs that they are not very tenable. Speaking generally, the koklass is a grey-bodied bird, with the centre of the under-parts chestnut, reaching right up to the neck in front, and sometimes extending backwards on it. The grey feathers are streaked with black, and either of these colours may predominate at the expense of the others, while the characters may be combined in different ways. The birds range nearly all along the Himalayas, and the local variations may be thus summed up.

In the North-west, where the typical *macrolopha* is the form found, there is as much grey as black on the body-feathers. In Nepalese birds, the so-called *P. nepalensis*, which is figured separately by Hume, though he himself was inclined to treat them all as one species, the body-feathers have more black than grey, so that the bird looks much darker. In Kashmir birds, which have been distinguished as *P. biddulphi*, the peculiarity consists in an extension of the chestnut colour on the sides of the neck. Even in the "Fauna of British India" the *P. castanea* of Kafiristan, Yassin, Chitral, and Swat, which is very little known, is kept distinct, and I followed this in my own book on Indian game-birds; but it really seems rather absurd to keep it separate, its only distinction being the great exaggeration of the chestnut round the neck and along the flanks.

The hens show very much less difference, though the Nepalese specimens run to a good deal of chestnut in the tail; there is nothing about their brown variegated plumage to attract attention, except the very pointed shape of the feathers all over, which is common to both sexes of the koklass, as well as the pointed tail. The hens have a very short blunt crest, and are not spurred like the cocks, which also have longer legs.

Another noticeable and characteristic point about koklass is that they have no bare skin about the face like most pheasants, and that their wings are unusually long and pointed for pheasants, more like a dove's, in fact. Connected with this is the great speed in flight, which exceeds that of any of our other pheasants.

¼

PUCRASIA MACROLOPHA

The cocks weigh from about two to nearly three pounds, the hens up to two, the Nepal race being smaller than the typical one. The propensity of naturalists for species-niggling forces one to waste a good deal of space in describing variations; coming to more practical points, the koklass is generally reckoned the best bird both for shooting and for eating of all its tribe in India; indeed, Hume says that he "would rather have a good day after koklass in the middle of November, in some little wooded saucer-like valley or depression at 7,000 or 8,000 feet in the Himalayas, where two or three coveys have been marked by one's shikaris, than after any other bird in any other place." Besides such places as are here indicated, koklass, he says, also especially affect "some place in a gorge where a horizontal plateau is thrown out inside the gorge." The birds keep much to the same place, though moving up and down during the day, and should be worked with well-trained dogs and several beaters.

The birds keep to the wooded parts of the hills, and range up as far as these extend, but do not go lower than about 3,000 feet, preferring the lower to the higher elevations, and liking sloping ground and ravines, especially when the trees are oaks. They are found singly and in pairs as well as in coveys, the last being family parties; the pairs are generally to be found near each other.

In places where there is little underbrush, they will run before rising, but otherwise get on the wing, though not till closely approached and forced to rise. Their very rapid flight down hill calls for good shooting; dogs will often put them up into trees, but when disturbed by man they will fly far and pitch on the ground, where they sometimes roost, though their general habit is to roost in trees.

They sometimes croak or chuckle when rising, whence no doubt the name of *Koak* in Kulu; in Kashmir they are called *Plas*. The *Koklass* or *Pokras* note, preceded by a *kok kok*, is the crow, and in dark shady woods in the interior they will answer any loud noise with it, though it is usually a morning and evening call.

They may be found scratching for insects in rhododendron

covert, and also eat moss, seeds, and flowers, and especially buds and leaves, but not grain ; they are not easy subjects for captivity, and are seldom kept, whence no doubt it comes that there is, apparently, no description of the cock's display extant. They do not breed lower than 5,000 feet, but may do so at twice that elevation, laying, with practically no preparation, on the ground under a rock or root, or in cover, buff eggs which fall into two types, the finely and uniformly speckled or the boldly blotched, the markings in both cases being reddish-brown ; the eggs also vary much in size, but average about two inches long. Some are much like those of our British black-game. Nine is the usual number, and May the usual laying month. Both cock and hen keep with the brood, and the young cocks get their colour in the first season, the young being well grown by September.

Cheer Pheasant.

**Catreus wallichi.* *Cheer*, Hindustani.

The Cheer Pheasant, although his colours have none of that brilliancy which one associates with pheasants, especially those with the typical long pointed tail which he exhibits in perfection, having this appendage sometimes two feet long, is nevertheless a very recognizable bird, not only among our Indian game-birds, but anywhere, for he is the only pheasant known which combines a long pointed tail with a crest also long and pointed; and the female, though shorter in both tail and crest, yet has them enough developed to be recognizable.

Although there is plenty of difference in detail between the cock and hen Cheer Pheasants, their general appearance is far more alike than that of the two sexes of pheasants in general, both showing black, grey, white, buff, and brown in their plumage ; the most noticeable differences are at the two ends, the cock having a plain dirty-white neck below his drab cap, while the hen, with the same head colouring, has the neck below the throat more black than white, though the colours are mixed ; her tail, also,

* *Phasianus* on plate.

PUCRASIA NIPALENSIS

though exhibiting the same colours as the male's, is not so distinctly marked, the cock's tail being boldly banded with black-and-tan on a bright buff ground, and forming a very noticeable feature in his appearance. Cock Cheer are much larger than hens, weighing about three pounds and often more, while the hens weigh two to two and a half; they look about as big as our cock pheasants at home, and this is the only one of our common hill pheasants, rightly so-called or not, which will strike anyone as closely like the home bird, in spite of its dull colour.

Its note, however, is, like its plumage, very unlike the common pheasant's, being a sort of song, rendered by Wilson as "*chir-a-pir, chir-a-pir, chir chir, chirwa, chirwa*"; but the tune varies, and there is a good deal of it to be heard, for hens crow as well as cocks, and in dull weather at any time in the day, though the usual calling-time is daybreak and dusk.

The cocks have spurs, and presumably they fight, for they are excessively spiteful in their demeanour to people when in captivity—more so, I think, than any other species; and they have considerable power in their strong bills, which they use for grubbing up roots, which are their favourite food, though they also partake of the other usual articles of pheasant diet, with the exception of herbage, for which they do not care.

Although distributed all along the Himalayas—to which range it is confined—and a common bird, the cheer is not to be found everywhere, its requirements being somewhat special. Although, like our pheasants generally, it ascends the hills in hot weather and descends in winter, it does not go above 10,000 feet or come down below 4,000, nor go outside the wooded regions. Even here Cheer are local, and the special grounds for them are, according to Hume, "the Dangs or precipitous places, so common in many parts of the interior; not vast bare cliffs, but a whole congeries of little cliffs one above the other, each perhaps from fifteen to thirty feet high, broken up by ledges, on which a man could barely walk, but thickly set with grass and bushes, and out of which grow up stunted trees, and from which hang down curious skeins of grey roots and mighty garlands of creepers." By waiting at the foot of such a place good shots may be got as the birds are driven down from above, but they come down extremely fast, apparently closing their wings and steering by their tails; while if hit and not killed they will run for miles at times. In thin tree-growth on the hillside they are hard to get unless bayed by dogs, at which, in out-of-the-way places, says Hume, they will chuckle or crow, with erected feathers, from the bough they have taken to, till they can be potted. Possibly this antipathy to dogs, like their fearless spitefulness to man when confined, indicates that they assist each other against vermin, for they are most companionable birds, except in the breeding-season, associating in coveys of up to fifteen in number, and these lots remaining about the same favourite place from one year's end to another, even if some are shot. They are great runners and skulkers when the grass is long and gives them a chance, and do not fly far at a time. In fact, they are essentially ground-birds, and seldom even roost on trees, but "jug" like partridges on the ground.

Cheer generally breed between 4,000 and 8,000 feet, preferably in May and at the foot of one of their favourite "Dangs" scratching a slight hole and laying small eggs for their size, not larger than a common fowl's, and dirty-white or pale-greyish, with a few rusty spots in most cases. The cock as well as the hen looks after the brood. The native name expressing the characteristic note is the most widely used, but in the hills north of Mussoorie is replaced by *Bunchil* or *Herril*, while in Chamba and Kullu *Chaman* is this bird's title.

Stone's Pheasant.

Phasianus elegans.

Stone's pheasant is one of the numerous subspecies of our common European pheasant (*P. colchicus*), and the hen is not noticeably distinct from the female of that bird; the cock also is likely to be considered the same on a casual view, but it really rather approaches the Chinese ring-necked race (*P. torquatus*), having the same lavender back and patches on the wings. There is, however, no white ring round the neck, and the breast is not coppery-gold as in the common pheasant and ring-neck, but dark

PHASIANUS WALLICHII

green. Thus a broad band of richly glossed dark colour runs right down the under-parts, completely separating the brassy-chestnut of the flanks. From Mrs. Hume's pheasant, the only one similar in form found with us, the present bird is distinguished at once by not having the white bars on the wing.

Stone's pheasant is found at an elevation of about 5,000 feet in the Northern Shan States, as well as at Momien in Yunnan, and in Szechuen. Its habits present nothing worthy of special mention; in Yunnan it was found frequenting grassy hills; and it may be remarked that the pheasants of this type naturally affect grass, reeds, and scrub-jungle, not high forest.

The Chinese ring-neck, which, subject to local variations, ranges from Kobdo to Canton, is the best all-round sporting bird in the world; it is now thoroughly mixed up with the original pheasant brought by the ancients into Europe from Asia Minor, and most English pheasants show traces of intermixture with it. It has been established also in places so wide apart as Oregon, New Zealand, Samoa, and St. Helena, and is often imported alive into Calcutta, where I have known an unmated hen lay and try to hatch her eggs. Such an adaptable bird is well worthy of introduction almost anywhere, and might be tried in the Ceylon hills and in the Nilgiris.

Mrs. Hume's Pheasant.

Calophasis humiæ. *Loe-nin-koi*, Manipur.

Anyone coming across this pheasant is likely at once to notice its resemblance to our familiar species at home, to which, indeed, it is nearly allied, though not nearly so closely as is Stone's pheasant. It may be distinguished from that bird by the two white bars on the wings and by the white edgings to the feathers of the lower back, which in some specimens conceal the dark bases, so that these would show a conspicuous white patch in that region which would be very noticeable when the bird was on the wing.

Such white-backed specimens are to be found in the Ruby Mines district in Burma, and some writers consider them as a distinguishable species, named *Calophasis burmanicus.* The typical form with the lower back having a variegated colouring of steel-blue with white feather-borders is the Manipur bird, and it is this that Hume discovered and named after his wife—a way of commemorating oneself (by giving the lady's married surname instead of her Christian name), which is, unfortunately, not unique in the annals of descriptive ornithology. This was in 1881, and Hume could only get two specimens, both of them males; but though few have since come to hand, the female is now known, and the Shan States, as well as Burma, have been added to the range of the species.

The hen, in her brown mottled plumage, has nothing distinctive about her appearance but the chestnut white-tipped outer tail-feathers, and fortunately these are just what would be conspicuous in flight; her tail is shorter than that of an English hen pheasant, though the cock's is quite up to the usual cock-pheasant's standard of length, but grey in ground colour instead of the olive-brown seen in the home cock-pheasant's tail.

In Manipur these birds are found inhabiting hill-forests, and range from 2,500 to 8,000 feet; they extend, according to Hume, "right through the Kamhow territory into Eastern Looshai, and North-west Independent Burma."

The Burmese and Shan States race, which was described by Oates as distinct in 1898, seems to be similar in habits, also frequenting wooded hills. Although I was the first to draw attention to the distinction between the two races, I did not, and do not now, consider them as distinct species, the characters being liable to variation; and I have always thought that the describing of a new species is an act requiring justification, not one to be proud of.

As an example of the futility of species-splitting, I may mention that two male specimens of this pheasant in the Indian Museum, obtained respectively by Lieutenant H. H. Turner in the Chin Hills, and by Lieutenant H. Wood in Upper Burma, agreed with the Manipur form in having the rump blue with narrow white edgings. As Hume's birds were trapped, and few have seen the species wild, Lieutenant Turner's notes are worth quoting; they appeared in the *Journal of the Asiatic Society* for

1900. He says : "I had left my camp, which was pitched about six miles from Fort White, on the evening of March 6 . . . and was returning along the road (the Fort White—Kalemyo road), when glancing down the *khud* I saw something grey disappearing in the long grass just below me. I immediately started to go after it, when I saw what appeared to me a light blue streak just disappearing. I immediately fired, but it was with faint hopes that I walked up to the spot, as not only did I think the bird had disappeared before I shot, but I had just at the moment of shooting slipped. I was, therefore, very much delighted when I saw the blue streak tumbling down the *khud* below me. I immediately went after him and secured him ; as I was descending the original grey bird, which was evidently the female, got up and flew a short distance. I walked her up, and my dog again put her up ; unfortunately, owing to the thick jungle, I was unable to get a shot. Walking on, however, I put up another, whether a cock or hen I could not say, as it was already dusk. I fired, but the bird flew away, and although I believe it dropped, I could not find it. These birds, when I saw them, were feeding among the dry leaves which littered the ground.

"The next evening I tried the upper side of the road and put up several (four at least) of these same birds out of some long grass on a steep hillside. I only managed to get one long shot, which was not successful. I again tried the next morning, and was successful in bagging another ; my dog put it up on our right, and flying very low through the bushes it crossed just in front of me. . . . The hill on which I obtained these specimens was between 4,000 and 5,000 feet high."

Lady Amherst's Pheasant.

Chrysolophus amherstiæ. *Seng-ky,* Chinese.

The striking contrast of satiny-green and white in the cock Amherst pheasant's plumage would be quite sufficient for identification even without its structural peculiarities of wig or frill of long rounded feathers, and extravagant length of tail which may reach over a yard, although the bird himself is barely as big as a hen common pheasant. The frill and long centre tail-feathers are both white marked with black, and are set off by the narrow red crest, red border to the straw-yellow rump, and red tips to the long tail-coverts ; the rest of the plumage is mostly green, but white below the breast. When displaying, the cock expands his tail and frill sideways, and always attracts attention at the Zoo when thus showing off; in fact, many people must know this bird by sight, although it is not yet, after many years' breeding in captivity, anything like so well known as its only near relative, the gold pheasant (*Chrysolophus pictus*).

The Amherst hen, though a plain-looking brown bird without trimmings, is strikingly marked off from our other hen pheasants by the bold cross barring of her upper plumage and neck; she has also chestnut eyebrows and a bare livid patch round the eye. Yearling cocks may be distinguished by the whitish tint of their napes and centre tail-feathers, and green gloss on the crown. Like its ally, the gold pheasant, this is a Chinese bird, but ranges to Tibet and reaches our territory also, though this has only been known in recent years. In his "Manual of the Game Birds of India," vol. ii, published in 1899, Oates mentions that he had seen a skin of a bird of this species, a cock, which had been shot, either in the Myitkyina or in the Bhamo district, on the frontier between Burma and China, by one of the officers engaged in the settlement of the frontier in question. Then, in 1905, Mr. E. Comber recorded in the *Journal of the Bombay Natural History Society* that the Society had "lately received the skin of an adult male specimen in full plumage of Lady Amherst's pheasant from Lieutenant W. W. Von Someran, who shot it at a height of about 9,000 feet near Sadon in the Myitkyina district of Upper Burma." The donor had stated about the habits of the birds that they lived at elevations of 8,000 feet or over, and he had never seen a bird below this; and that they appeared to be common over the frontier on the hills of the Chinese side.

The habits of the bird in China are thus described by Père David, "Lady Amherst's pheasant lives, the whole year round, in the highest jungle-covered hills of Western Szechuen,

Yunnan, Kouycheou, and the highest hills of Eastern Tibet. It especially frequents the clumps of wild bamboos which grow at an altitude of 2,000 to 3,000 metres, and the shoots of these are its favourite food; indeed, it is from this that its Chinese name of Séng-ky (shoot-fowl) is derived. . . . . In the wild state it shows a very jealous disposition and will not allow the golden pheasant, its only possible rival, to approach the locality in which it resides; and so one never meets these two brilliantly coloured pheasants on the same hill or in the same valley." Another clerical authority, quoted by Hume, says that Amherst pheasants, when they find springes baited with grain laid for them, are said by the Chinese to try to sweep the corn away with their huge tails so as to feed safely on it. This sounds rather a tall statement, as Hume evidently thought, but it is quite possible that the Amherst cock, one of the most irritable birds in a very peppery family, may, in his anger at being kept from coveted food by an obstruction which he fears, may play round the snare with expanded sweeping tail as he would round a hen; for this species, like probably most birds, assumes more or less the so-called courting attitude under strong emotion such as anger. Of course any native onlooker at this performance, if it occurs, would naturally credit the bird with an intelligent motive. If some corn were actually swept away in this manner, it would indeed be probable that the bird would learn to act intelligently in the asserted direction. The birds, as above remarked, breed freely in captivity, and their eggs are buff.

Indian Crimson Tragopan.

Ceriornis satyra. *Munal*, Hindustani.

The wonderfully rich plumage of the cock crimson tragopan, whose red under-parts spotted with white, and the similar speckling on his marbled brown back, make him look like a glorified guinea-fowl, is a certain and striking distinction of his species; the hen is a brown bird, the plumage on close inspection being seen to be a grizzly pepper and ginger mixture, with more of the dark colour above and more of the buff below, but without definite markings of any size; she is quite easy to recognize, in spite of her sombre colour and absence of any crest or bare skin round the eye. Young cocks show some red on the neck in their first year, but do not come into colour till the next. The cock is horned, crested, and dewlapped, as is always the case with tragopans; but the crest lies flat and the light blue fleshy horns are generally concealed in it, while the dewlap is hardly visible as a rule, just showing a fold of the richest blue skin on the bare throat. The blue skin of the face is concealed by scanty black feathering; and in having the face thus feathered this species is unique among tragopans. Although, like our other well-known tragopan, this species is often called argus, it is no more an argus pheasant than it is a peacock; indeed, it can hardly be called a pheasant at all, being, like the monal, a member of a separate group in the family, and quite as near the partridges as the pheasants proper. The tail is somewhat hen-like, not long, and slightly folded, and the general appearance is bulky and fowl-like, though the legs and toes are rather long and slender, and the bill particularly small. The bird is a large one, weighing about four pounds in the case of cocks; the hens are noticeably smaller and do not weigh nearly three pounds.

The crimson tragopan is confined to the Eastern Himalayas, seldom straying west of the Alaknanda Valley in Garhwal, in which state and in Kumaun it is known as *Lungi;* it is well known as far east as Bhutan, where its names are *Omo* and *Bap*, the Lepcha name in Sikkim being *Tarr hyak*. It used to be common near Darjeeling.

Like tragopans generally, it is a true forest bird and seldom seen, for it does not come out on to the grass slopes above the forest as the monal so frequently does; though, like that species, it shifts its ground according to season, keeping near the limits of woodland in summer, and descending in winter as low as 6,000 feet. It likes thick cover, and is especially fond of that afforded by ringal, especially where water is at hand. It is more of a tree-bird than pheasants generally, not only taking refuge in trees from enemies and roosting on them at night, but judging from the habits of captured specimens, keeping a good deal in them at all times, and no doubt feeding on the buds, berries, and leaves, since leaves, especially of aromatic kinds, and wild fruit, form a portion of the food, as well as bamboo-shoots, insects, and bulbs; though in confinement it will eat grain, it does not seem to seek it in a wild state.

Although eggs have been taken in Kumaun in May, not much is on record about the breeding of this bird in the wild state, no doubt because people naturally expect such birds to nest on the ground, whereas evidence obtained from birds kept in captivity shows that they are really tree-breeders. Mr. St. Quintin, who has paid particular attention to tragopans and kept three out of the five known species, finds they require elevated nesting-sites, such as old wood pigeons' nests and platforms put up in trees, which they line with a few twigs. A hen of this very species even made a scanty nest of her own with spruce twigs and branches, so that in looking for tragopans' nests one's motto evidently ought to be "Excelsior." The eggs are larger than ordinary Indian fowls' eggs, and not unlike them except for a few pale dull markings of a lilac tint. They take twenty-nine days to hatch.

The chicks are uniform reddish-brown above, not striped, and have the wing-feathers showing when hatched; they perch at once, and can fly in a few days. This looks as if they might spend some of their early life aloft; perhaps the hen feeds them, as the cock does her when courting. This same courtship of the cock is very curious; he has two quite distinct displays, an unusual trait in any bird. The most commonly seen is a sideway one, the bird flattening himself out sideways, as it were, by expanding the feathers of one side of the body above and below, much as the common pheasant does. In this way the white spots become as conspicuous as possible, but there is no change in the face. In the full display, which is very rarely seen, the bird squats down on his heels with head erect and plumage puffed out, flaps his wings with a convulsive movement, showing off the intense red on the pinion-joints, and makes a noise like a motor-car starting. At the same time, with jerks of the head, the dewlap is let down and expands, not vertically like a turkey's, but horizontally, forming a bib as large as a lady's palm, of the most intense blue in the middle, and pure azure at the sides, which are marked with large oval scarlet spots. The horns should be also displayed at this time, but I hardly saw them when I witnessed the display myself. This is a frontal display, but the hen never seems to be anywhere where she is wanted at the time. There appears to be usually but one hen with a cock, and he seems more gentle with her than typical pheasants. Her alarm note is much like the quack of a duck; the cock is usually silent, but in the pairing season calls with a bleat like a young lamb, and also, but for only two or three days in each season, according to Mr. Barnby Smith, who has carefully studied this species in confinement, gives out a weird, far-reaching, moaning call like *oo-ah, oo-ah*, apparently as a challenge. Cocks can be called up by imitating them, but are even then very wary and hard to shoot; in fact, it is very difficult to get a sight of tragopans at any time, and the peculiarities of their display have been made out from captive birds. As a general rule, unless they can be hustled out of the ringal cover by dogs and made to rise, they afford very little sport, for when seen in the open, as they rarely are, they break away on foot if possible and give only a snap-shot. They are often not better eating than ordinary fowls, so that on the whole, though most fascinating to the naturalist, they do not figure prominently on the game list.

Temminck's Tragopan.

Tragopan temmincki. *Oua-oua-ky*, Chinese.

Temminck's tragopan, as may be judged from its alternative name of Chinese crimson tragopan, is a bird whose most conspicuous colour is red, as in our eastern Indian species; but the Chinese bird is a perfectly distinct species, not a mere local race, although the two are undoubtedly far nearer to each other than either is to any of the few other tragopans known.

The characteristic points of the Chinese bird are, first, the bareness of the face, which permits the bright blue colour of the skin to appear, and makes the bird in life conspicuously different from the Indian bird with its black-feathered countenance; and secondly, the fact that the plumage is spotted, not with white, but with grey, and that these light spots have not the black borders which so throw up the pearl spangling of the crimson tragopan of India. The spots are also larger in Temminck's tragopan, especially on the under-surface, where as much grey shows as red, the feathers being practically grey with broad red borders.

The bib, as expanded during courtship, is of apparently the same colour in both tragopans, being blue with a row of scarlet patches down each side; at least that is what I have noted, having seen each species display.

I can give no criterion for distinguishing the hen of this bird from that of the Indian crimson tragopan; but as no two tragopans have been found living together in our borders as yet, the problem of separating these is not likely to arise.

As it occurs on the Mishmi Hills, the Chinese crimson tragopan was long suspected to be a likely resident in our borders, and this suspicion became certainty in 1908, when Mr. E. C. S. Baker reported to the Bombay Natural History Society on two specimens which had been "shot by Mr. W. Scott, Civil Officer of the Sadon Hill Tracts, on the Panseng Pass at a height of 9,000 feet. Mr. Scott in a forwarding letter described the bird's call as "one single, high note, not unlike a cat's mew."

It is the south-western and central parts of China that are the best known home of this species, but it appears to be, according to Mr. Baker, very common above 8,000 feet on the Mishmi, Dafla, and Abu Hills; in Sadya it is found on the high ranges within only a dozen miles of the frontier police posts.

Père David, writing of its habits in China, says it is not common anywhere; it lives alone on bush-covered hills and rarely comes out of its cover, where it feeds on seeds, fruits, and leaves. He says its very sonorous cry can be represented by the syllable *oua* twice repeated, whence one of its Chinese names; the syllable *ky* means fowl, as in the two other names, *Ko* or *Kiao-ky* (horned fowl) or *Sin-tsiou-ky* (starred fowl). He says it is a much esteemed game bird, all the more so because it is so scarce and can only be captured by a trap or springe.

In captivity in Europe it is as well known as the Indian

species, or at any rate used to be, but of late years I have only seen the Indian crimson bird at the Zoo, though the only other Chinese tragopan known, Cabot's or the buff-breasted (*T. caboti*), has been exhibited of late and been not uncommon in the bird trade. In captivity the Temminck's tragopan shows the same tendency to nest high up as the Indian crimson species; the eggs are cream or buff colour closely speckled with brown.

Western Tragopan.

Tragopan melanocephalus. *Jewar*, Garhwal.

The "Simla argus," as this tragopan is sometimes called, the crimson bird being the "Sikkim argus"—both wrongly, for as I said before, they are not at all like argus pheasants—is sufficiently like its Eastern relative to be recognized as a close kinsman at once; there are the same white spots, the same general size and form, and the same red on the neck and pinions, while the ground-colour of the back is of a similar mottled brown. But the under-parts are very different, being nearly all black in ground-colour, thus enhancing the guinea-fowl effect, while the face is quite bare and bright-red, although the bib is said to show both red and blue, and is probably similar, when fully expanded, to that of the better-known species.

The hen is more of a true pepper-and-salt grizzle, with less rufous in the tint, and on the under-parts is distinctly spotted with white; her hues are altogether colder than those of the crimson bird's female, as one would expect from the sparseness of the red colouring in her mate, which would really be better called the black tragopan, from his dominant colour.

The young cock, as in the other species, first shows his colour on the neck; he is said not to come into full colour till the third year. This species runs a little larger than the crimson bird; it is found from the ridge between the Kaltor and Billing rivers in native Garhwal, on the east, all along the hills as far as Hazara, being known in the north-west as *Sing-monal.* As the crimson tragopan is also called *Monal* in Nepal, it seems that natives group the great pheasant-partridges, as one may call these birds, and the true monals together. In Kullu, Mandi, and Suket there are different names for the sexes, the cock being *Jigurana* and the hen *Budal;* the Chamba name is *Falgur*, and that used in Bashahr is *Jaghi*.

Unlike so many representative species, the two tragopans do not range up to each others' boundaries, for, says Hume, from the ridge in Garhwal above-mentioned, "for some four days' march you meet with neither species. In this interval there are three high ranges to cross that divide the Bhilling Rand Valley from that of the Bangar Rand, this latter from the Mandagni Valley, and this latter again from that of the Alaknanda." How it is the birds have left this considerable bit of neutral ground untenanted appears never to have been explained, and the problem would be well worth solving.

Like the crimson tragopan this species is essentially a woodlander; it feeds chiefly on leaves, especially of box, oak, ringal, and a privet-like shrub; it also likes berries, especially that of the *Dekha* of Kullu, and takes insects, acorns, and grubs as well, while in captivity it eats grain. Though shifting its ground more or less according to season, and ascending in the spring to near the forest limit, it often remains in forests with plenty of snow on the ground, being able to find its food in the trees. It is a shy bird, avoiding human habitations, and seldom seen even by natives, while, though it becomes tame very quickly in captivity, it seems rarely to be exported, so that its intimate habits and display are apparently unknown. The wild alarm note is a repeated bleat like a lamb's or kid's, and the spring call is a loud version of the same; no doubt there are really two notes as in the crimson tragopan. Where not disturbed, these birds may be seen at times feeding in open patches in the forests along with monal, and are easily shot when treed by dogs; but persecution makes them very wary, and at the best of times a pot-shot on the ground or in a tree is all that can be got. They hide themselves with great skill, and when "treed" watch the sportsman and shoot off as soon as discovered before proper aim can be taken. They generally keep in straggling parties, and are often found alone.

The eggs have rarely been taken, owing probably to the

assumption that birds of this kind must be ground-breeders. Those that have been taken are dull freckled buff; six were in the clutch, and May was the month in which they were taken, at the western limit of the bird's range in Hazara. They were on the ground, in a spot where a landslip had carried away a bit of pine-forest, covered with small second growth of bushes and shrubs; the nest was a rough structure of grass and sticks. No doubt if old pigeons' and squirrels' nests are investigated in this tragopan's haunts, the eggs will be more easily found.

Grey-breasted or Blyth's Tragopan.

Tragopan blythi. *Sansaria*, Assamese.

The grey-breasted tragopan is distinguished at once from our other species by the spotless smoke-grey of the under-parts, although the upper plumage is mottled much as in our other species, and the neck is of the same red; the fleshy horns are also of the usual blue, but the face-skin is bright rich yellow, bordered with green where it ends on the throat. Although I have seen the bird alive, I have never witnessed its display, so cannot give the colour of the bib, which of course can only be properly seen in the live bird.

Although rather smaller the hen is very similar to that of the crimson tragopan, but the under plumage is less rich in tint, and there is more of the black peppering in the grizzled brown of the upper plumage.

Little is known of this beautiful bird, although it was described by Jerdon as long ago as 1869; it is best known from the Naga Hills, though it also ranges into Manipur and Cachar, and has been reported from the Daphla Hills also. The Nagas, who know it by the name of *gnu*, are in the habit of catching it "by laying a line of snares across a ravine which they are known to frequent, and then, with a large circle of beaters, driving the birds down to them. They go as quietly as possible, so as not to frighten the birds sufficiently to make them take flight, as if not much alarmed they prefer running." This bird's habits are, in fact, evidently much the same as those of other tragopans, the group being as much alike in their ways as they are in their general appearance, although the species are so well characterized and distinct. The cry is evidently some sort of a bleat as in the other species, as it is said to be expressed in the syllable "*ak*."

The habitat of the bird is high jungle, and it does not seem to range lower than 5,000 feet, while going up to the tops of its native hills. The breeding season is said to be in April, and three or four eggs to form the full clutch; these eggs appear to be of the buff, brown-spotted type, normal in the group.

In Cachar Mr. E. G. S. Baker once watched, in April, a pair, of which the male was "busy courting the hen who refused all advances. They behaved exactly like domestic fowls, and the cock kept running round the female with trailing wing." This, however, judging from what has been seen of other better-known tragopans, would only be the simpler and commoner form of display, so that the full show posture evidently remains to be recorded.

One of the first specimens known was sent home alive to the Zoo, and they have had several others since; in fact, the bird seems to have become better known in captivity than in its wild state.

Monaul.

Lophophorus refulgens. *Munal*, Hindustani.

An American naturalist has well said that this gorgeous bird reminds one of a humming-bird enlarged to the size of a fowl; and really this does give one some idea of the remarkable appearance of this glory of our hills, for only among the humming-birds do we find such brilliant green and copper as clothe the cock monaul's head and neck, while the purple and blue of his back and wings are only second in brilliancy to the tints further forward. As the bird flies off, however, two more hues are particularly striking, the snow-white patch in the middle of the back, completely hidden in repose by the wings, and the rich chestnut of the short broad partridge-like tail. In fact, in spite of the brilliant colours and peacock-like crest of the male, which have given him the name of *Nil-mor* and *Jungli-mor* in Kashmir, there is something very partridge-like about the bird, and to call him the Impeyan *Pheasant,* as is often done, is rather an abuse of terms, for, although a member of the pheasant family, he is no more a pheasant than he is a jungle fowl or a peacock, but, with his few relatives, stands alone as a type.

The hen, in her mottled-brown plumage, is just like a giant partridge; her only distinctive marks are the bare blue eye-patch she shares with the cock, and the pure white of the throat. Yearling cocks may be at once picked out on the wing by the patch of plain buff which foreshadows the snow-white escutcheon they will bear when in full plumage, which is not till next year, and even second-year birds have the seventh quill brown. The monaul is a heavy, bulky bird, weighing about four and a half pounds in the case of the cock, the hen being about half a pound less. It carries a great amount of breast-meat, and tastes much like a turkey, at any rate during autumn and winter; the thigh sinews run to bone, and need drawing like a turkey's. The monaul is confined to the Himalayas, and is seldom found lower than 5,000 feet even in winter, while in summer it ranges up to the forest limit and even above it, some old males climbing nearly to the snow-line. However, it is, generally speaking, a forest bird, and so usually roosts in trees at night, besides often alighting on them in the day-time when disturbed.

It is a strong-flying wary bird, more of a flyer than a runner, and quite ready to cross a wide valley on the wing when surprised. Its call, which is somewhat like that of the peewit at home, but a whistled instead of a mewing call, is a source of annoyance to sportsmen after big game in the heights, for of course the beasts attend to the warning. Both cock and hen call similarly, and the note is quite unlike that of any other of our game-birds.

Monaul are, like the family generally, mixed feeders, but they specialize on underground food—roots and grubs—and hoe these up with their powerful bills: they rarely scratch like the pheasants and fowls, being able to do the work of unearthing food with the bill alone as a rule. In the wild state they do not care for corn, but will eat it in captivity, especially wheat; but any-one keeping them should always supply chopped roots as well.

They are not very sociable, and old males are often found alone; their spurs are short, and one does not hear about their fighting in a wild state, though in captivity a strong male will hunt a weaker one to death, and I have known a vicious youngster to completely scalp a hen. But, on the whole, they are gentle, quiet birds compared with the excitable pheasants. The display of the cock is curious—he begins by bending down his head and expanding the turquoise eye-patch; then he sets out his wings without fully expanding them, and raises and spreads his tail, thus showing all his top-colour at once. When thus at full show he parades with mincing gait round the hen, now and then hopping in a way strangely out of place for so heavy and dignified a bird. He often has but one mate, but in localities where the species is common several may fall to his lot. In fact, Wilson, the "Mountaineer" so well known in Indian sporting literature for his unrivalled accounts of our Indian hill game-birds, found that by rigidly preserving hens he could market male skins of this species and the western tragopan by hundreds yearly without decreasing the stock, so that polygamy is quite a workable arrangement for the species, although Mr. St. Quintin, who has bred it in confinement in England, finds that the cock looks after and broods the chicks as well as the hen. But this may have been due to isolation; the general impression in India is that the hen only tends the brood.

Owing to the value of its jewelled plumage the bird has been liable to be much poached by natives, who capture it with nooses and dead-falls, all of which devices ought to be strictly forbidden, as they are fatal to hens as well as cocks. To the legitimate exploitation of the males no reasonable person should object, but these game-birds need careful protection, and if the natives' poaching propensities could be directed to the destruction of the numerous vermin of India a great point would be gained. In this connection it should be mentioned that the hawk-eagle is an inveterate foe of this bird and of tragopans, while no doubt the marten accounts for a good many.

Monauls breed in late spring, the hen making a "scrape" under a root or rock, and laying seldom more than five eggs; they can be first found in May, and may easily be mistaken for those of a turkey, but are slightly larger than the eggs laid by Indian turkeys at any rate.

The native name *Munal*, with the feminine *Munali*, is especially used in the Central Himalayas; in Kulu the male is distinguished as *Nil* and the female as *Karari;* in Kashmir the sex-appellations are *Lont* for the cock and *Hami* for the hen; *Ratnal* and *Ratkap* are used in the North-west Himalayas, while the Lepchas and Bhutias call the bird *Fo-dong* and *Cham-dong* respectively, and *Dafia* is the name in Nepal, recalling the term *Datiya* in Kumaun and Garhwal.

There is a certain tendency to variation in the plumage of the male monaul, and in some cases this has led to some unsatisfactory species being named; a form with blue instead of copper-red on the neck has been called *Lophophorus mantoui*, and in several books a variety is called the BRONZE-BACKED MONAUL, and credited with being the true *Lophophorus impeyanus*, whereas it always used to be supposed that it was the typical form which was named after Impey. This variety, for I personally cannot swallow it as a species, and natives say it is only a casual variation, has only been found in Chamba, where the common form is well known. It is distinguished from this bird by having much more metallic gloss on the plumage, there being no white patch on the back, but purple all the way down, while the green of the throat spreads all over the under-parts, which are intense velvet-black in the typical bird. As no hens ever turn up, and as birds only differing in colour invariably interbreed and do not themselves recognize a difference of species, I really think naturalists have been too much in a hurry in giving specific rank to this freak; for it so happens that the pheasant family are particularly apt to produce well-marked and natural-looking colour-variations, of which the black-winged peacock, also once ranked as a species, is a striking example.[1]

---

[1] Since writing the above, I find that Mr. C. W. Beebe has published (*Zoologica*, vol. i, p. 272) his conviction that the Chamba monaul is "unquestionably a mutation, sport, or abnormal variation."

Crestless or Sclater's Monaul.

Lophophorus sclateri.

Although this fine bird has not yet occurred in Indian limits, it is very likely to be found to do so, since it inhabits the Mishmi Hill, like the Chinese crimson tragopan, now definitely established as an inhabitant of our Empire. If met with it is extremely easy to recognize, for, in spite of a general resemblance to the common monaul, it has two very marked points of distinction, one at each end—the absence of the crest, combined with a peculiar frizzling of the scalp-plumage, and the white tip to the tail. The white patch on the back also extends right down to the root of the tail, not being separated from it by a dark glossy area as in the common monaul ; and this in the case of a captive bird, which is likely to have a broken tail and probably a damaged scalp as well, will no doubt prove the best distinction.

The hen bird, since the question of crest does not come in, is naturally more like the hen common monaul, but even in her case there is a clear and easily-seen distinction ; for she also has a white-tipped tail, and if this mark, owing to damage, be not available for recognition, the noticeable light area on the lower back will show a difference from the common monaul's female.

The first specimen of this bird on record was seen by Jerdon in 1869, and, though it was in bad feather, he, with his great knowledge of birds, divined it was probably a novelty, and proposed the scientific name it now bears. The bird was then living at Shillong, healthy, though in damaged condition; it ultimately reached the London Zoo. As only a few specimens have turned up since, brought down by Mishmis and Abors to the annual fair at Sadiya, there is hardly anything to say about this bird, one of the most gorgeous in existence.

Blood Pheasant.

Ithagenes cruentus. *Chilimé,* Nepalese.

In spite of his striking plumage of slate-grey, pale-green, and carmine, the cock of our Alpine blood pheasant looks, on the whole, more like a partridge, having a short tail and only weighing a little over the pound; while the hen, being brown all over, would certainly be called a partridge by anyone who did not know her mate. Her bright red legs and red eye-patch, which she has in common with the cock, are distinctive points, as also is the fine pencilling of black over the brown plumage, which has no striking markings.

Young cocks are said to assume a duller edition of the masculine plumage when half-grown; they have no spurs, but their elders are most plentifully provided in that respect, and may have up to nine spurs on the two legs.

The Bhutias, who call the bird *Some* or *Semo,* credit it with growing a new spur every year, but this is at least doubtful, and the bird is so rarely kept in captivity that opportunities for observation have been wanting. One pair reached the London Zoo a few years back, and I was struck with their essentially partridge-like appearance. Their importer, Mr. W. Frost, told me that they were spiteful with other birds, and backed each other up, the hen waiting on an elevated spot till the cock ran a bird under her, when she would spring on it and do her share of the mauling.

That the bird should be seldom kept alive is not remarkable, for it is not often even shot; it is purely Himalayan—though very similar species occur outside our limits—and always keeps high up near the snows, but affects cover, not open rocky spots like the snow-partridge. Pine forests and mountain bamboo clumps are favourite haunts, and here the birds scratch for food like fowls, and are nearly equally omnivorous in their tastes. But, like most of our game-birds, they specialize somewhat in food; they do not eat bulbs, and do eat pine tops and juniper berries, especially in winter and spring, for they remain all the year at high elevations. As they do not range lower than 10,000 feet, their haunts are liable to be snowed up, but in addition to the food they get from the conifers, they seem to burrow in the snow for either subsistence or shelter; for they have been taken at 12,000 feet in January.

They perch freely at all times when alarmed, but fly little and generally run to cover when startled; the alarm-note is "ship, ship," and a scattered covey is piped together by a long squealing call. The covey varies in number from ten to twenty birds, and in winter packs of up to a hundred may be found. Not much is really known about these birds, which seem to have their haunts very much to themselves. Even the eggs have not been taken, but these must be laid pretty early, for young ones are about in May; and Jerdon got the half-grown birds on the Singhallala spur west of Darjeeling in September, a locality unusually near the plains for this species.

It may be gathered from what has been said about its running habits that this bird is not of the sport-showing description; but, occurring as it does where other game is scarce, it is useful for food if one is hard up for meat. But it is an uncertain article of diet, for though it has been found excellent eating in September, after feeding on berries, leaves, and seeds, a diet of coniferous vegetables reduces it to a condition of rankness and toughness that requires a really keen appetite to overcome; so that it is a bird to be left alone as long as even village fowls can be procured.

Mountain Quail.

Ophrysia superciliosa.

Anyone lucky enough to start this curious little bird in shooting in the hills might recognize it by its tail, which is far bigger than in any quail-like bird, whether true quail, bush quail or button quail, being in fact three inches long, while the bird itself is little bigger than the common grey quail.

A true quail it certainly is not; some call it a pigmy pheasant, and it may be that, if the blood pheasant is fairly called a pheasant, for to that bird it seems to be allied. Like it, it has long soft plumage and red legs; but it has no spurs, and the colour of the sexes, though different, does not present the striking contrast of the cock and hen blood pheasants. The cock mountain quail is grey, narrowly streaked with black along the edges of the feathers, and the hen brown, also variegated with black markings, but in her case these are broader and occupy the centre of the feather. There is, in fact, nothing in her colour to attract attention, but the cock is noticeable for the rather striking black-and-white colouring of his head, and of the feathers under the tail. Both have red bills, brightest in the cock; but in some apparently the legs and bill may be yellow, as in the first recorded specimens, which were in the Earl of Derby's private zoo in 1846.

It was not certain that these came from India, but nobody has found the bird anywhere else; and even there it has only rarely turned up, always in the hills, and generally in winter. Less than a dozen specimens, in fact, are on record, and all these have been got near Mussoorie or Naini Tal. Hume suggested that they may have come "from the better-wooded south-eastern portions of Chinese Tibet," which little-known region might certainly furnish novelties. But the bird does not look at all a wanderer; its wings are small even for a bird of this family, none of which have pinions adapted for lengthened flight. The common grey quail is the best provided in this respect, and that has wings of four inches or more from the pinion to the tip—the usual way of measuring a bird's wing, as it can be done in a skin made up as usual with closed wings; the mountain quail, although larger than the common quail, shows a wing of barely more than three-and-a-half inches measured in this way.

It may be that the birds obtained represent some of the last survivors of a declining species; such species must always be in existence, and may no doubt disappear without record, for extinction of course goes on, as it did before the advent of man with his much-abused destructive habits, from natural causes.

In the Naini Tal Tarai, for instance, there exists a large weaver bird or baya, the *Ploceus megarhynchus* of Hume, of which very few specimens have ever been obtained; yet this is the brightest-coloured as well as the largest of the Indian bayas, the cock in breeding-dress being nearly all yellow, on the throat and belly as well as the breast and cap. This may be a declining form; but against the theory of imminent extinction in the case of the mountain quail, and in favour of that of migration, may be set the dates of the latter bird's occurrence, which are almost all in the winter months. Thus it has occurred near Mussoorie in November, 1865, and close to Naini Tal in December, 1876. In November, 1867, however, a number appeared at Jerepani,

♀ BAMBISICOLA FYTCHII.

and some of these were still there in June of the following year, but were not seen later.

Like a partridge, this bird is found in coveys, as well as in pairs or alone; it is extremely hard to put up out of the long grass or other low cover in which it lives, finding its food in the grass-seeds, and only taking a short slow flight when disturbed. Its presence, however, is often betrayed by its whistling call, which is quite peculiar. Being a hard bird to shoot and poor eating, there is not much inducement to go after it, and for the last thirty-eight years none have been seen or heard of either in India or anywhere else.

Bamboo Partridge.

Bambusicola fytchii.

The general impression made by this bird may be judged of by the fact that an escaped specimen in England some years ago figured in a sporting paper as a hybrid between a partridge and a pheasant; it is, indeed, a partridge in size, but its tail in length and form rather recalls the pheasant type. The plumage is sober and partridge-like, and the same in both sexes; the distinctive points about it are the chestnut spots on the brown back, and the diamond or heart-shaped black markings on the belly. The legs are grey, spurred in the cock, and often in the hen as well.

The bamboo partridge was first discovered by Dr. Anderson at Ponsee, in Yunnan; here it frequented old rice land on hillsides at 3,000 feet. It is now known to inhabit Manipur, the Kachyen Hills, and in our territory, Upper Burma and the hill ranges west to Assam. This being so, it at first seems strange that a good-sized bird like this should have first been made known from outside, when it occurs even near Shillong, but it is a very skulking bird, and difficult to flush. Besides bamboo-jungle, it haunts long grass and heavy forest jungle; it is strictly a hill bird keeping above 2,000 feet.

It perches freely and roosts in trees at night; and on rising in the morning will come out into open spaces. It is not an abundant bird, and generally found in pairs ; its call, heard in spring, is unmusical and loud, like *che-ke keree.* The nesting season is said to be May and June, and the eggs brownish-buff laid in a nest on the ground in or under a tuft of grass. The weight is about twelve ounces in the cock and a couple less in the hen ; the plumage is exceedingly variable in detail, but the points given above, in conjunction with the length of tail, which may reach five inches, make this bird easily distinguishable from other local partridges, especially the woodland species, all of which except this one have very short tails.

Himalayan Snow-cock.

Tetraogallus himalayensis. *Ramchukar*, Hindustani.

The home of this great grey partridge, as big as a small goose, is the rocky but grassed slopes between the forests and the snows on the Himalayas ; its eastern limit is Kumaun, and it ranges on the west through Afghanistan, where it is called *kabk-i-dara,* to Central Asia.

Seen on its native heath, or rather turf, it looks, from its grey colour and orange legs, large size, and rather awkward gait, very like a goose ; it also has other goose-like habits, feeding mostly on grass, though now and then scratching up a tasty bulb, and being eminently sociable, several old birds being seen together with a number of chicks ; while the sentinel perched on a stone ready to give warning to the pack is eminently reminiscent of the ways of wild geese. When on the wing, these birds fly well and often high, frequently crossing from one ridge to another, or travelling a mile at a time, and they are particularly conspicuous when in flight, owing to their pinion-quills being white except at the tips, while they keep up a continual whistle while flying. They habitually feed up-hill, walk slowly, and never run far ; in fact they are not built for much sprinting, being thickset, short-legged birds and very heavy, the cock weighing up to six and a half pounds, and normally about five. The hen is not nearly so large, but still weighs between three or four pounds, and except for having

Ceriornis melanocephalus.

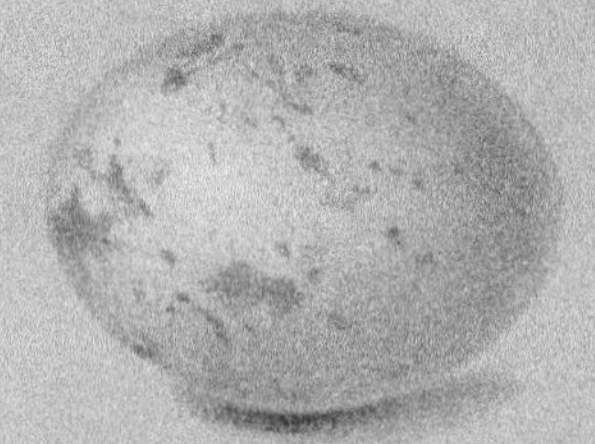

Porzana akool.

Excalfactoria sinensis.

Perdix hodgsoniæ

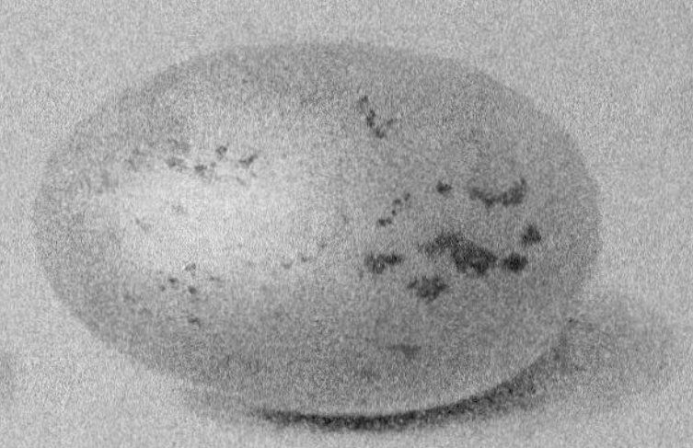

Pterocles senegalus

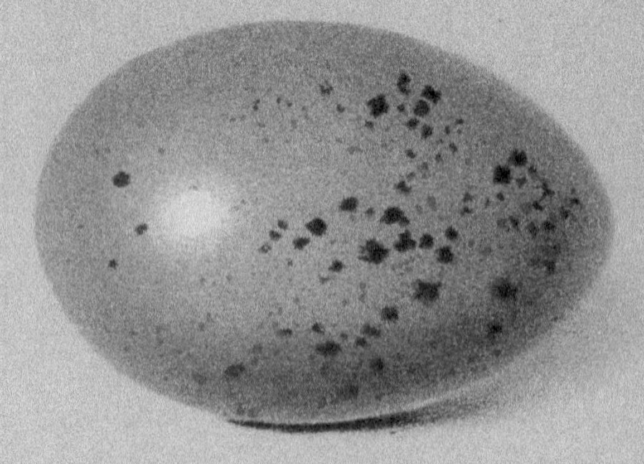

Tetraogallus himalayensis.

A. W. Strutt. Del.

F. Waller. Chromo Lith. 12 Hatton Garden London

no spurs, is just like the cock, both having the same chestnut-edged white bib and white breast, and chestnut streaks on the grey ground of the wings and sides.

Snow-cocks, often somewhat absurdly called snow-pheasants, for they are most obvious partridges in everything except size, avoid cover of any sort, but they are rather partial to rocks, and roost on the shelves of precipices at night. They like feeding on spots where sheep have been folded at night earlier in the year, as the grazing is better in such places; and on cold, dull, and wet days keep on the feed all day, though warm bright weather makes them sluggish and disinclined to leave their rocky perches except at morning and evening. They are, indeed, essentially birds of the cold bleak heights, and few remain to breed on the Indian side of the Gangetic section of the mountains, the majority here apparently crossing the snows to nest in Chinese Tibet, though in Kunawar they are common at all seasons.

In September they appear between the woods and the snow, and as winter draws on the heavy falls drive them down to any open hills they may find in the forest belt. Their migration seems to be made at night, and in mild winters hardly any come down; 7,000 feet is about the limit in any case. Once settled on a hill, they stay till the end of March, and each pack has its own location, to which it appears to return every year. They will feed on young sprouting corn very readily, and eat other herbage besides grass, but only visit isolated patches of cultivation.

Generally speaking, they dislike a nearer approach than about eighty yards, and though they will merely walk off at first if approached from below, an intruder from above will make them take wing almost at once; while their vigilant sentries see to it that no advance is made unnoted. Generally speaking, therefore, they need a rifle to bring them to book, and as their ground is also frequented by burrhel and tahr, many people find them rather a nuisance than otherwise, since when out with a rifle men prefer the four-footed game, and the alarm-whistle of the birds startles these. Moreover, although such fine big birds, and usually very fat, they are indifferent eating at best, and often positively nasty, no doubt on account of some herbs or roots they eat. All birds with this attribute of occasional unpleasantness, by the way, ought to be drawn as soon as killed, as this often prevents the tainting of the flesh by the food which may have been eaten; and in any case some natives will eat them, so that shooting them is not by any means wanton destruction.

Their chief enemy appears to be the golden eagle, but as he prefers, according to Wilson's excellent account of this species, to take his game sitting, and the snow-cock naturally does not wait for this, but flies off before his tyrant stoops, he does not often get one. But this may only apply to the young eagle, the ring-tail as Wilson calls it, from the banded appearance of the tail, which has a white base in the young; no doubt the older birds learn by practice to catch their prey flying, and in fact I have read somewhere a description of such a chase in which the eagle used his advantage of height to drop on the flying snow-cock before the victim had got up full speed.

The comparatively few birds which breed on the Indian side of the Himalayas nest from 12,000 feet upwards to the snows, making a "scrape" in some spot well sheltered from rain. The eggs are not unlike turkeys' eggs, but darker and greener in the ground-colour, an olive or brownish stone-colour in fact, with fine brown spots. Five is the usual clutch, and when more are seen it is to be suspected that two pairs have "pooled" their broods, though many pairs separate and bring up their young by themselves in the usual manner of partridges. The eggs are generally laid by the end of May, but sometimes not till early in July. This conspicuous bird naturally has many names: *Huinwal* in Kumaun, *Kubuk* or *Gourkagu* in Kashmir, *Leep* in Kulu, and *Kullu, Lupu,* or *Baera* in Western Nepal, though the bird is not actually found in Nepal itself; the Mussoori hillmen's name, *Jer-moonal,* implies a recognition of the relationship of this great partridge to the short-tailed so-called pheasants of the tragopan and the monaul groups.

Tibetan Snow-cock.

Tetraogallus tibetanus. *Hrak-pa*, Bhutanese.

Although found in our territory from Sikkim to Ladak, it is only at the highest elevations that this species of snow-cock is to be found, and even on the wing it may be noticed as something different from the ordinary kind, not showing the conspicuous white on the pinion-quills, which are dark with white tips instead of the reverse. Close at hand the differences in plumage are even more striking, for the grey colour of the upper-parts only extends across the breast, the belly being white with black streaks. Here again, then, there is a reversal of colours in the two species, the common snow-cock having a grey belly and white breast, slightly barred transversely with back. Moreover, although the legs are of much the same colour in both species, the bare skin near the eye is red in the present bird, and yellow in the other.

The Tibetan snow-cock is a much smaller bird than the Himalayan, the cock barely equalling the hen of that bird in size; and in the Tibetan species the hen is not much smaller than the cock.

The real home of this desolation-loving bird is the northern side of the mountains between India and Tibet, and it is generally distributed over the latter country, extending to Turkestan westwards, and east to Kansu and southern Koko-Nor. In the Himalayas it is seldom found lower than 15,000 feet, and occurs up to 19,000. A sure find for it appears to be the Sanpo Pass, where it is particularly common. Scully found it abundant near there in 1874, having seen hundreds in one day; he found them excellent eating and not very shy.

According to Prjevalski, who observed it in its north-eastern haunts, this is a lively, noisy bird, with several calls—a note, uttered at rest, much like a common hen's, varied by a snipe-like whistle, a *click, click*, when alighting, and a *goooo, gooo* when settling down on the ground; while it has a distinct whistle for reassembling a scattered brood.

He considered the birds very wild, and found them good runners; but both he and Scully noticed that they would not stand an approach from above so readily as from below, flying in that case instead of running, and thus in this point resembling their Himalayan relatives.

This snow-cock does not seem to nest on our side of the hills, and not very much is known about its breeding anywhere. But Prjevalsky found them pairing in April, and came across young in August, some no bigger than quails, and others full-sized; so that, here again like the Himalayan species, they must lay at different times.

The eggs appear to be greenish-white with dark spots. Both parents lead the brood of from five to ten young, and when these are fledged, the whole take wing together and do not settle till they have put a ravine or valley between themselves and their pursuers. They moult in August, and even in September were not fat, though natives said that they did become so in autumn, so that towards winter they would probably be in good condition for the table.

Grey Partridge.

Francolinus pondicerianus. *Titar*, Hindustani.

The grey partridge, which is one of the sub-group of partridges known as francolins, is *the* partridge of India, and to it the name *titar* especially applies, though it is sometimes called *gora* or *safed titar*, to distinguish it, no doubt, from another very well-known francolin, the black partridge. It is not really grey any more than the so-called grey partridge which takes its place in Europe, but brown with pale cross-pencillings, not very unlike that bird, above; but below it is decidedly different, showing none of the grey on the breast which the European common partridge (*Perdix perdix*) has, nor the "horseshoe" on the lower chest; the under-parts in our Indian bird are barred with fine rather sparse dark cross-lines on a pale buff ground. The throat is unmarked, and outlined by a rather imperfect black necklace. In the common partridge of India there is not even the small sex difference that occurs in the European bird's plumage, the cock being only distinguished by his spurs, which are well developed. The legs are red, but not bright as in the chukor or the "red-leg" at home.

All these points are easily to be studied by any newcomer to India before he goes out to shoot, for this partridge, being the favourite fighting bird among sporting natives, is constantly to be seen in cages everywhere, and its characteristic call, *ká, ká, kateetur, kateetur*, as it is well rendered by Hume, is the first game-bird's note one is likely to hear other than the degenerate utterances of the domesticated descendants of the mallard and jungle-fowl and the cooing of the blue pigeons.

Almost wherever one goes in India one is likely to find this bird, and it is also found in the north of Ceylon, where it is called *Oussa-watuwa*, but is not an inhabitant of Burma, though in the opposite direction it is found outside our limits to the Gulf. It is absent in swampy districts and heavy jungle, and does not occur south of Bombay on the Malabar coast-line, nor is it found in Lower Bengal, being a bird of dry, warm soils, low cover and cultivation. But, although it is to some extent a percher, taking readily to trees when alarmed, and often roosting in them, it can do without such cover as well as without cultivation, and exist, if the ground be broken, in practically desert localities, as in the Sind hills. It does not go high in the hills anywhere, a couple of thousand feet being its limit.

It is a bold bird, not only feeding on ploughed and stubble fields, but on roads, and visiting threshing-floors in the early mornings; in fact, it hangs about villages so much that it shares the unsavoury reputation of the hare and the village fowl. Grain of all sorts it gladly eats, and also takes grass and seeds, young leaves and insects, especially white ants, breaking up a nest of these being an excellent way to attract partridges. Even when it has been living on irreproachable diet, however, this partridge is poor and dry compared with his savoury relative in Britain, and although he flies more smartly and strongly, has a great objection to doing so, and will run so persistently that to follow him is only missing chances at quail and hares, which are more certain shots; though in some places, as in heavy grain crops on cloddy soil, the little skulkers can be made to rise willy-nilly, and then furnish good enough sport. They can also be treed by any dog which will hunt, and shot in this way, but in the ordinary way are only subjects for chance shots and not a regular object of pursuit.

They are found both in pairs and in coveys, the latter presumably being family parties, the cocks being far too quarrelsome to live together; and they are prolific birds, for though nine is more than the usual number of the creamy-white eggs, they breed twice a year, at any rate in many cases, the spring nesting beginning in February, and the later about August. The nesting habits vary curiously; the nest may be practically non-existent, the eggs being laid on the ground, or it may be a hollow in a tussock or under a bush, more or less lined with grass, or even, a most remarkable site, made in the branches of a thick shrub a yard off the ground. The Bengali name of this bird is *Khyr*, the Tamil *Kondari; Kawunzu* is the Telugu appellation.

Swamp Partridge.

Francolinus gularis. *Bhil-titar*, Cachari.

The khyah, as this fine partridge is called by the Bengalis, is, unlike most game-birds, essentially a bird of swampy and alluvial soil, overgrown by high grass and cane; but even where it is commonly found along with other partridges, as it is in some places, it is a very distinctive bird. It is as big as a jungle hen, looks all its size on account of its long legs, and has a very smart appearance; its upper plumage is much like that of the common grey partridge, but the under-parts, with their well-marked broad longitudinal white streaking on a brown ground, contrasting with the rich rust-red of the throat, are most characteristic, and make one wonder how people could ever have mixed this bird up with the chukor, with which it has nothing in common except being of good size for a partridge and having red legs, though these are not bright in tint. The male has sharp spurs, but this is the only sex distinction except his slightly larger size.

The swamps of the Tarai, and the low-lying lands along the courses of the Ganges, Megna, and Brahmaputra, and the lower reaches of their tributaries, are the habitat of this bird; Cachar is its eastern limit, and it is found as far west as Pilibhit. Considering its tastes in locality, it is curious that it does not occur in the Sundarbans, and that it sometimes is found on land of as much as 4,000 feet elevation. Some of its haunts are so low that it is driven by floods to take to the trees in the rains, or to leave its home altogether and resort to cultivation or bush-jungle. It is rare to find it in grass low enough to go after it on foot, and when on cultivated land the birds have a sentry posted on some bush. They are found in pairs, threes or coveys, and are more noisy and quarrelsome, if anything, than their smaller relative the grey partridge, whose call their own resembles, but with the last syllable cut off; evidences of their desperate battles are found in the honourable scars which adorn the breast of so many specimens, but these veterans are but dry eating, as may well be supposed. Their asserted enmity to the black partridge has been doubted, on the ground that the two species may in some places be flushed out of the same grass cover, but it is probable that in the breeding-season this large and fierce bird is a serious enemy to other partridges of smaller size, though there may be a truce at other times. At any rate, those who value the fine black partridge should have an eye on the "grass chukor" till he has been proved innocent; though as a rule his preferences in the matter of habitat must make him harmless in most cases.

If worked for from elephants, the kyah gives very good sport, but its flustering, cackling rise is rather trying at first to the nerves of behemoth if not to his rider. From the high cover it affects the breeding of this species is naturally not much under observation; the eggs seem to be five in number, and are slightly darker than those of the grey partridge, and likewise differ in being sparingly marked with pale brown or lilac at the large end, having in this, it must be admitted, a slight resemblance to the real chukor's eggs. The nest is on the ground among high grass, and has been found in April.

Besides *Bhil-titur*, this bird is called *Bun* and *Jungli-titur* in Hindustani; *Kaijah* is another Bengali name, and the Assamese is *Koera; Koi* is also used in Assam.

Black Partridge.

Francolinus vulgaris. *Kala titar*, Hindustani.

The black partridge, which is the original and typical francolin, is at once distinguished from all other Indian partridges by the prevalence of black in his colour; his white cheeks and chestnut collar, and the beautiful variegation of white in pencilling on the tail and rump and spotting on the sides, make him one of the most beautiful partridges known, and, unlike the generality of our partridges, very distinguishable even from his own hen. In her, the markings are in the less contrasting tones of brown and buff, and differ somewhat in detail, the under-parts in old hens, at any rate, being pencilled, not spotted, while the collar is reduced to a patch of chestnut at the nape; but this is quite enough to distinguish her from our other brown partridges, none of which have this chestnut nape-patch. The legs are orange, spurred in the cock. The weight of a cock black partridge is up to twenty ounces, though some are only half that weight, as the birds vary greatly in size; hens run two or three ounces less than cocks.

The marked distinction between the sexes is indicated by the Hindustani name *Kais-titur* for the female, *Kala*, of course, especially indicating the male; the Garhwal name *Tetra* rather recalls the Greek *tetrax* for some game-bird, but is more probably related to the Hindustani *Titur;* in Manipur, the farthest point east at which the bird is found, it is called *Vrembi.* It is one of the few Indian birds of non-migratory habit which extend to Europe; at any rate, it is still found in Cyprus as well as throughout Asia Minor, but is now extinct in the countries bordering the Mediterranean on the north, though formerly found even as far as Spain. As the classical ancients were as much given to introducing game as we are, it seems possible that the bird was in Greece, Spain and Italy, and the islands only an exotic after all, so that it is less surprising that it has failed to maintain itself. Even in India, though such a well-known bird, it has its definite limits; it is only found in the northern provinces, and does not descend into Kattywar or below Orissa. Nor does it inhabit the hills above 7,000 feet, and to this elevation it only attains viâ the river valleys. It is also a bird of cover and cultivation, eschewing desert tracts, and affecting especially the sides of rivers where there is a thick growth of grass and tamarisk, as well as thin jungle, even scrub on very dry ground; away from some sort of sheltering wild vegetation it is not to be found except as a straggler in most cases, though it is willing to haunt sugar-cane fields.

In spite of its preference for cover, it is essentially a ground-bird, and, though in some localities it may take to a tree to call, it generally, even under the circumstances of delivering its morning message to the world, uses an ant-hill, fence, or rock as a pulpit. The call is harsh and metallic, of about half a dozen syllables, of which various renderings exist both in English and Hindustani, for there is something about the note which impels many people to try and put it into words. Hume says "Be quick, pay your debts" is about the best English version. The call is most heard in breeding-time and winter.

The black partridge—although it is almost always in pairs, the family coveys only keeping together for a very short time—is very common in some localities, though, alas! all too readily shot out. It is the best of Indian partridges as a game bird, and with the next species enjoys a somewhat similar status to the grey partridge at home, while if not quite so good as that bird on the table, it is nothing to grumble at as a game course. It feeds, like the grey partridge, on insects, shoots, and seeds and grain, and is not always to be depended on for scrupulousness in diet when near villages, though not in this or in any way so low-caste a bird as the grey partridge. Blacks are not nearly so quarrelsome, and far less addicted to running, so that they afford really satisfactory sport; they can be shot, according to the height of the cover, either on foot or from an elephant, and, in Hume's time, at the beginning of the eighties, fifty brace a day might be bagged by one gun, while far higher numbers are on record.

This valuable bird, however, can be and has been worked for more than it is worth; although it lays as many as a dozen eggs, it generally fails to raise more than a quarter of such a brood till they are even three parts grown, and this is put down to persecution by vermin, which are allowed to work their will unchecked in India. Proper game preservation and due consideration in shooting—it might be as well to limit the bag to the easily distinguishable cocks—ought to make this bird as abundant as our home partridge in all suitable localities.

The nest is very well hidden in crops or tamarisk or grass jungle, and is of the usual partridge type—a scrape and a wisp; the eggs, most often laid towards the end of June, are glossy and spotless, and are of a stone or fawn tint tinged with green or brown. It would very likely be a good plan to put some of these eggs in the nests of the grey partridge, removing the original ones, and see if this hardy but less valuable bird would rear the young of its betters, as this plan has often succeeded with game and other birds; but to find nests of the black partridge the aid of a good dog is very often requisite.

Painted Partridge.

Francolinus pictus. *Kakhera kodi,* Telugu.

From the usual use of the word "painted" in characterizing birds, one would expect this species to be at least as handsome as the black, its near ally, which indeed is also called *Kala titar* by the Mahrattas; but as a matter of fact it is not nearly so showy a bird, wanting the white cheeks and chestnut collar, though the face and throat are chestnut, and having the white spotting below so developed at the expense of the black background that the general effect is light below variegated with dark, and the bird on the whole is rather more like the hen black partridge than the cock, though much purer in its colours.

The hen of the present species is much like her mate—who has no spurs—but may be distinguished by the throat being white, not chestnut like the cheeks, and by the light barring on the lower back being coarse and buff, as in the similar marking in the hen black partridge; in the cocks the rump-pencilling is narrow and pure white in both kinds. That the birds are closely related they themselves recognize, for cross-pairing takes place

HYBRID— ½ FRANCOLINUS PHAYRII.

BETWEEN FRANCOLINUS PICTUS & FRANCOLINUS VULGARIS.

now and then on their borders, resulting in hybrids. It would be interesting to know which way these are bred; theoretically, the handsomer and better-armed male of the black partridge ought to be able to elope with the ladies of the present species, which is moreover a decidedly smaller bird, but questions like these can never be settled theoretically, and observation often results in a surprise.

The mention above of the frontier of these two birds coinciding illustrates the fact that the painted partridge is a southern Indian bird, which ranges even to Ceylon, though curiously enough it is not found in Mysore, or south of Coimbatore or of Bombay on the Malabar Coast. In Ceylon it is confined to some hills in the Newera Eliya district, and is not found anywhere outside it. The name "southern francolin" well expresses its position, though "painted" quite well describes it in comparison with the ordinary grey partridge, if not with its handsome dark northern cousin.

Although so nearly related to this bird its ways differ considerably in detail. For instance, though the relationship is recognizable even in the notes, the calls of the two birds are not identical, that of the present one being rendered as *Chee-kee-kerray*; it also calls even earlier in the morning, and generally from a tree, in which the singer and his mate have probably passed the night, for this bird is far more of a percher than the last species or than most of our partridges, and is commonly to be found in trees in the morning and evening. The cocks, by the way, call very late as well as very early.

The sort of localities which the black partridge affects are not so much favoured by the southern francolin, which is more partial to dry soil, and less fond of jungle; in fact, cultivated fields, if well supplied with trees, are a pretty sure resort for these birds, as is also scrub jungle on rocky ground; but they also haunt sugar-cane fields, and are in fact pretty easily suited, though in many districts very local.

Although more given to running than the northern francolin, they are nearly as good both for shooting and eating, and can claim the only place near these birds in point of all-round excellence among our lowland small game. They feed on the same sort of food as black partridges, and, like them, are not beyond suspicion if shot near villages; they also go in pairs—the coveys in this case too not keeping together long—and are not quarrelsome. The eggs of this bird are most often laid in August, and are deposited in similar places to those of the northern bird, and about as hard to find; but they are smaller in size, duller in surface and fewer in number. In colour they are some shade of cream or drab, without spots. The young have a "peculiar cricket-like chirrup" which they begin as soon as they are hatched. This species is pretty uniform in size, and only weighs from about eight and a half to twelve and a half ounces, never approaching some blacks.

Eastern or Burmese Francolin.

*Francolinus chinensis.** *Kha*, Burmese.

This third member of this beautiful spotted group of partridges is to a certain extent intermediate between the two Indian species; though not so large as the northern bird it is bigger than the southern, and the cock has more black in his under-surface colouring than the latter, and also possesses spurs. He has no chestnut collar, however, and in his head-colouring he is very distinct from either, and perhaps even handsomer; for the face and throat are white with a bridle-like cheek-stripe of black. This marking is repeated on the head of the hen in buff and brown, otherwise she is very like the hen of the southern francolin or painted partridge.

The Burmese francolin is not generally distributed over Burma, but only in the Thoungyen and Irrawaddy valleys and in Karennee and Tonghoo. In Karennee and down the Irrawaddy to Prome it is a common bird, and being a noisy one, especially in the breeding season, June and July, its presence is obvious enough. The call is of the same character as that of black partridge, but has a style of its own, and is

---

**phayrii* on plate.

rendered by Wardlaw Ramsay as *kuk, kuk, kuich, ká, ká*. When calling it perches on a stump or branch. In Karennee it frequents the slopes of rocky hills, and in Pegu scrub jungle, waste land and open places in forests, and is partial to bamboo jungle, sometimes coming into paddy fields after harvest, though it avoids open country as a rule. It is fond, however, of the thick cover of deserted clearings. In the Thayetmyo district it is very common and appears more or less independent of water, which is here scarce; Oates suggests that this is probably due to its food consisting largely of buds and shoots, as well as insects. It does not seem to be much of a grain feeder, and is rarely seen in stubble.

It will only rise when driven by beaters from its cover, and even then drops again as soon as possible, though it is a strong flyer and gives a sporting shot. It is good eating, according to Schomburgk, who met with it in Siam, and was told it roosted in trees. It is also found commonly in South China, Hainan, and Hongkong.

It seems to breed chiefly in Upper Pegu with us, and lays from four to eight eggs, which are very like those of the black partridge, unspotted and of a greenish cream, buff, or stone-colour. The nests found have been on the ground, but Schomburgk was told that in Siam these birds nested in trees. There also he found it frequenting rice-fields and pasture-grounds in flocks, which does not agree with its habits in our territory, where it does not much affect cultivation and is not social, though many may be found in one place; this is perhaps what Schomburgk means, and at any rate Swinhoe expressly says that in Hongkong it is solitary and does not associate in coveys, which is the character of these spotted francolins generally. Swinhoe also says the flesh is insipid, which also is more in accordance with what is known of the other species, though much of course in matters of flavour depends on local or seasonal circumstances; no doubt grain-fed birds might be as good as our home partridges, as Schomburgk says these are.

Chukor.

Caccabis chucar. *Chukar*, Hindustani.

The chukor, a very familiar bird in the hilly and mountainous parts of Northern India, is distinguished from all other partridges by having no pencilling or streaking upon the upper plumage, which is also much greyer than in any of the others, in fact, often more grey than brown; the colour varies according to situation, the greyest birds being found in dry districts with but little cover, and the brownest in places where there is plenty of moisture and shelter. This is, in fact, the most versatile of all partridges in its choice of a habitat; as Hume says, "In one place it faces a noonday temperature of 150° F., in another braves a cold, about daybreak, little above zero: here it thrives where the annual rainfall exceeds 100 inches, and there flourishes where it is practically *nil*."

In its red beak and feet, the legs armed in the cock with a lumpy apology for a spur, its black necklace round a white (sometimes buff) throat, and the handsome vertical bars of black and brown on its blue-grey flanks, this bird at once recalls to mind its near relative, the "Frenchman," at home. It is a good big bird as partridges go, though very variable, hens being about a pound, while cocks may be half as heavy again. In the Himalayas it ranges up to 16,000 feet, being found in Ladakh, but is also found in the comparatively rich country of the southern hills, and ranges down to the Punjab hills and the barren rocky Mekran coast; but it is not found east of the Indus in Sind, and does not extend to Sikkim. In Kashmir it is known as *Kau-kau*, in Chamba as *Chukru*.

It is a sedentary bird, not wandering much from its chosen haunts wherever they may be situated; and though grassy hills are a favourite resort, rocky, bushy ravines will also hold chukor, and they like the vicinity of cultivation, and often glean in the cornfields in autumn; jungle they entirely avoid. In winter quite large packs, even up to a hundred in number, may be met with, where the birds are plentiful, but in any case coveys are the rule at this season, though when breeding they pair off as usual. The cocks fight furiously in the breeding season, and are often kept by natives as fighting birds.

In high desert places they are very wild if they have been at all shot at, and give hardly any sport, but in the lower Himalayan hills they are far more easily accounted for, though they have the "Frenchman's" trick of running. They fly more strongly and sharply than our common partridges, and come downhill at a great pace if pushed up by dogs; but they do not like rising after one good flight, and in such a case will often lie well when walked up. The native name, universally adopted by Europeans, is simply the bird's call, and it is very liberal with its note, especially when a covey has been scattered. The food of chukor consists of grain and other seeds, and often of insects; they also take much gravel for digestive purposes. Old birds are dry and tough, but young ones good in autumn if hung; they may be distinguished by having black instead of red bills.

Chukor breed at any height in the Himalayas over 4,000 feet; in the Punjab Salt Range and the lower Himalayas they lay in April, but higher up the nesting may be deferred till three months later. The nest may be a mere scratching, or a pad of grass and leaves, and contain as many as fourteen eggs, or as few as half that number; they are yellowish white, peppered or spotted with brown, much like the eggs of the French partridge, in fact, as one would expect. Outside India the chukor is found in Tibet and the Thian-shan range, and east to China, west to Aden in one direction and Cyprus in the other; while the so-called Greek partridge (*Caccabis saxatilis*) of Eastern Europe only differs by having the face with a little more black, reaching to the corner of the mouth, in the eastern bird there is merely a continuation of the necklace beyond the eye up to the nostrils. The difference is unimportant, and the two birds are obviously races of one species; while in the really distinct west European, French partridge (*Caccabis rufa*) there is the definite distinction of a fringe of black spots outside the necklace-band.

Seesee.

Ammoperdix bonhami. *Sisi*, Hindustani.

The sandy colour of the Seesee partridge, which much resembles that of the sand-grouse, whose desert dominion it invades, is a striking distinction from all our other partridges; indeed, of these only the chukor is ever likely to be found along with it, and the red legs and conspicuous necklace of this are very distinctive points, as well as its much larger size; the Seesee is a very small partridge, only weighing about half a pound.

As in the case of the sand-grouse, whose more pigeon-like build at once distinguishes it from the Seesee, the two sexes have plumage which, though producing the same sandy effect, and closely assimilating them to the soil, is yet different in detail. The hen's is obscurely pencilled, but has no distinct or striking markings; the cock is rather peppered, and has some distinct colour touches in the chestnut and black streaking on his sides, which rather recalls that of the chukor, but is longitudinal instead of transverse; the delicate grey on the head and throat, set off by a black eye stripe, are also distinctive of him, and he has a bright orange bill. When showing off to the hen, he stands erect, and puffs out his striped flank-feathers so as to make himself look rather like a goblet or a lady in a crinoline. He has no spurs, and is believed not to fight; but I should think that very doubtful.

Seesee are only found in the desert hills in the North-west, and even in such districts they prefer the barest rocky ground; though, as they feed on seeds and herbage, they not unfrequently come on to grassy places. But even scrub they usually avoid, for they need no cover, since all they have to do is to sit tight if they want to hide. Being great runners, and often over very bad ground at that, and having a trick of shooting straight downhill, they do not give much chance of a shot; in some places, however, they are remarkably tame; they give a rather harsh whistle when rising, but their characteristic call is the soft dissyllable which is imitated in their native name. They breed up to 4,000 feet elevation, laying usually amongst stones, with a little dry grass for a nest; eggs may be found as late as June, but generally some weeks earlier. The eggs are spotless cream colour, often nearly white, and the chicks when hatched are covered with creamy-buff down, and not striped as in the majority of young game birds; in fact, this species does pretty well conform throughout to the much overworked theory of protective coloration; and, indeed, as it is small and weak, and has not the powers of flight of the sand-grouse, one can see in its case why invisibility needs to be its main resource. It is found by some to be remarkably good eating; though Hume considered it inferior to a good chukor, and not of high quality. Outside India it ranges, if the country be dry enough, west to Persia, and a so-called species with but trifling distinctions (*Ammoperdix heyi*) is found on the confines of the Red Sea and Persian Gulf; but only these two, if they really are two, represent the particular type.

Tibetan Partridge.

Perdix hodgsoniæ. *Sakpha*, Tibetan.

Considering that our home grey partridge is generally speaking a lover of rich cultivated land, it is very surprising to find that the only one of our Indian partridges which can claim a close relationship with it is only to be looked for in the highest and most desolate parts of the Himalayas; though it is true that the first Indian specimen was shot on chukor ground in fields in the Bhagirathi. From chukor the present bird may be distinguished quite easily, according to Hume, who shot them on one of the high passes leading from the Indus to the Pangong lake. He noticed, he says, that "their whirring rise and flight were precisely those of the European bird and very different from that of the chukor." He was also led to search for them by hearing their calls, the remarkable similarity of which to that of the English partridge attracted his attention.

This similarity extends to the plumage taken as a whole, but there are plenty of differences in detail; most to be noticed is the white face and throat with well-marked black patch on each side, and the blotch of black, coalescing from black bars, on the part where the well-known horseshoe mark comes in the home partridge, when present. The weight of the Tibetan bird is a pound, and, like the home bird, it is excellent for the table. How it gets into good condition is rather a puzzle, for, according to Hume, its environment, like that of the snow partridge, which keeps fat on next to nothing, is not luxurious. He says, "The entire aspect of the hillside where these birds were found was dreary and desolate to a degree, no grass, no bushes, only here and there, fed by the melting snow above, little patches and streaks of mossy herbage, on which I suppose the birds must have been feeding." Prjevalsky, however, found the nearly allied race, *Perdix sifanica*, in rather less miserable surroundings in Alpine Kansu, where it inhabited rhododendron thickets. In Tibet its western limit seemed to be the Changchenmo valley; those found in our territory are derived apparently from the Chinese portion of Tibet, occurring in Kumaun and British Garhwal.

It has been found breeding near the Pangong lake, on ground where for 100 miles there was not even brushwood to break the monotony of rocky barrenness, and on the Oong Lung La Pass, leading from that lake valley to the Indus valley at an elevation of 16,430 feet, in this case among grass and low bushes. Ten eggs were in the clutch, but Prjevalsky says the Kansu bird lays fifteen. The Tibetan-taken egg which Hume obtained was of a glossy uniform drab, pale, but slightly tinged at each end, especially the larger, with reddish brown.

Snow-Partridge.

Lerwa nivicola. *Barf-ka-titar*, Hindustani.

When high up after mountain sheep and goats, on rocky ground near the snow line, one may start a covey of dark birds with conspicuous white patches on their wings, which spin away with grouse-like flight—evidently partridges of some kind, for there are no true grouse anywhere in India even in these Himalayan heights. These alpine partridges, the *Lerwa* of the Bhutanese, would be recognizable even if they lived among others, for their closely cross-pencilled plumage with chocolate

LERWA NIVICOLA.

belly, and brilliant red legs and bill, are striking when close at hand. Yet in their own haunts the birds are hard to see on the ground, in spite of there not being enough cover to hide a sparrow; for the snow partridge is true to its name, and if possible, will be on ground where the only vegetation is moss and the shortest of grass, interspersed with bare stone and snow patches.

On the scanty vegetation of these heights the birds contrive not only to live, but to get and keep fat; their usual weight is over a pound, and may reach nearly a pound and a half. In winter they are perforce driven to the lower hills, but always keep to their preference for barren spots, and manage to keep in touch with the snow, not descending below 7,000 feet.

Although not scarce birds and in some localities common enough for a hundred to be seen in a day's march, they are decidedly not generally distributed, and occur in localities often separated by a day's journey. Where rare, they are wild, but tamer where they are abundant, but in any case much shooting will naturally have the effect of making them less approachable; in favourable circumstances, they rank as some of the best sporting birds, and only the superior attractions of four-footed game cause them to be neglected. Although best known as a Himalayan species, this bird extends outside our boundaries as far as Western China; its western limit is Kashmir.

The call of these partridges is a loud harsh whistle, which they give out when approached, and they keep whistling when on the wing. They go in flocks or coveys except during the breeding-season, and even then sometimes several old birds may be found along with a number of young, as in the case of the snow-cock, which these birds resemble so closely in appearance and note when on the wing that unless it is possible to make out the great difference in size they are difficult to tell apart; but the snow-cock is found on rather different ground, and is wilder and takes longer flights.

They breed, as they live by preference, as near the snow-line as possible, on rather rough ground, on the ridges jutting out from the snow. The eggs are laid under a rock, apparently about the end of May, since chicks, according to Wilson, were first to be seen about the 20th of June. The eggs are large for the size of the bird, being bigger than those of the chukor, and are freckled with reddish brown on a dull white ground. About half a dozen chicks seem to occur in a brood ; they are mottled grey and black above, with three black stripes on the head, rufous under-parts, and black bills; a certain amount of black remains on the beak when they are full feathered. The old ones show great attachment to them, sometimes shamming lameness in the well-known partridge fashion, and at other times walking away before the intruders with piteous calls, while the little ones squat, or creep under the stones.

The snow-partridge is very good eating, and after keeping a few days resembles a grouse in flavour as much as in appearance. Besides the Hindustani equivalent of "snow-partridge," the bird is also called *Bhair titar*, *Ter titar*, and *Golabi titar*, while the Chamba name is *Biju* and the Kumaun one *Janguria* ; *Quoir monal* is given as the title in Garhwal.

Common Hill-Partridge.

Arboricola torqueola. *Peura*, Hindustani.

The characteristic peculiarities of this bird—the dumpy body, with short tucked-in tail, yet mounted on rather high legs, spurless but furnished on the toes with very long claws only slightly curved, are characteristic of hill-partridges in general, the most numerous but least interesting group of partridges found with us.

It has a greenish-brown back barred with black, and the sides are grey, streaked with chestnut and spotted with white. The head and breast differ in colour in the two sexes, but this is exceptional in the group, all the others having the sexes alike. In the cock peura the top of the head is chestnut, and the face and throat black slightly streaked with white, a band of which colour, unmixed with black, borders the grey breast above. The hen has a duller brown, black-streaked cap, and the ground-colour of the throat chestnut, though it also is marked with black ; the breast has a brownish tinge. In young birds of both sexes the white flank spots invade the breast also.

The name *Arboricola* (tree-haunter) is appropriate enough if it means a dweller among trees, for all these tree-partridges live in forest, and in forest on hills; but they do not live in the trees like tragopans, though they perch sometimes, but apparently not more than some of the other partridges, such as the painted species or southern francolin.

The range of the present species is wide, for in addition to the Himalayas it is found in the Naga Hills and in those north of Manipur. It is a bird of the middle hills, not generally going above 9,000 or below 5,000 feet, and its especial haunts are dark forest-clad ravines and gullies. It may, indeed, be found close up to the limits of vegetation, and is then more easily seen, so as to give an idea of abundance, but its real home is lower down; here, however, it is in its element as an accomplished skulker, and so is rarely seen. Dogs, however, will put it up, and its scent is so strong, says Hume, as to draw off the dogs from that of pheasants. Its flight is low, short, and very swift, and it must be hit by the snappiest of shots or not at all. It is practically omnivorous, eating both insects and leaves, seeds and berries, and may often be seen feeding near the various hill pheasants; open land and cultivation it avoids. As might be inferred from the length of the claws, it scratches for food a great deal.

The note is a soft whistle, either loud or low according to circumstances, only heard in spring, and easily imitated; indeed, so like is it to the whistle with which the shepherds call their flocks, according to Hume, that these simple hill-men believe that the birds are the abodes of the transmigrated souls of former colleagues, and in some places object to their being shot in consequence.

Peuras go generally in pairs or singly, though in autumn and winter coveys of half a dozen may collect. They are commonly only shot casually, according to the rather scanty opportunities they afford; about half a dozen a day may be thus picked up when after better game. As food they are dry, but go well enough in a stew. The eggs are said to be about half a dozen in number, and white; this, at any rate, is the usual colour of the egg in this very distinct group, which are all much alike in every way, chiefly differing in details of colour.

The present species, being a widespread bird, has several native names; that given above is in use in Nepal and Kumaun, as well as *Ban-titar*; the Lepcha name is *Kohum-pho*, and the Kangra one *Kaindal*; *Ram chukru* and *Roli* are used in Chamba.

Rufous-throated or Blyth's Hill-Partridge.

Arboricola rufogularis. *Pokhu*, Daphla Hills.

The rufous-throated hill-partridge, which in Kumaun shares the name of *peura* with the last species, is not unlike the hen of that bird, having the throat also chestnut variegated with black; but the brown back is scantily spotted with black, not profusely barred as in both sexes of the true peura, and the legs are red, whereas those of the common hill-partridge are grey, with only a tinge of red. In having the breast clear grey, not drab, this partridge resembles the male rather than the female of the common hill-partridge.

The rufous-throated hill-partridge, generally speaking, has the same range in longitude as the common species, though occurring in the Daphla Hills and those of Karennee and Tenasserim, where that is not found; but it ranges through a different zone in altitude, being a bird of the lower hills, and descending even to their foot. Above 6,000 feet it is not to be looked for, even in summer-time. In Tenasserim birds the feet are not so bright a red, the size is altogether larger, and they are generally without a black stripe, which in the Himalayan race is found at the termination of the reddish brown on the neck.

Besides keeping lower down, the red-throated hill-partridge is more sociable than the ordinary species, or, at any rate, forms larger coveys, according to Hume, but in other respects its ways are precisely similar. In the Daphla Hills it was noticed by Godwen-Austen to come down at night into the warm gullies, and feed upwards along the ridges, so that the natives were able to snare numbers by erecting little fences across their path, with openings set with bamboo-peel nooses, a method of capturing ground-game widely practised in eastern hills, and one that should be prohibited for its destructive results, in all districts where the natives do not really need wild creatures as food.

In Tenasserim Davison found these birds curiously tame; they would perch within a few feet of him, and sit there gazing and whistling, a proceeding strangely at variance with the usual retiring habits of this group. He says, by the way, that the calf is "a series of double whistles, commencing very soft and slow, but gradually becoming more and more rapid, and rising higher and higher, till at last the bird has to stop." This sounds as if the note were quite unlike the single whistle of the common hill-partridge, so that Hume was probably wrong in describing the calls as identical; in the call being some sort of a whistle all these hill-partridges agree, but differences in detail are just what might be expected. Blyth, by the way, found these birds rising solitarily in Tenasserim, and in winter at that, so the social habit is also liable to variation. The food of this species is seeds, small snails, and berries, and like the last, they are great scratchers among dead leaves. A heap of these has been found to form the nest, and the eggs are white, of a dirty shade, and very scantily and minutely speckled with grey. Four fresh ones were taken below Darjeeling, on July 4th, by Mandelli; but the full clutch may be larger, and no doubt earlier ones are to be found.

Arakan Hill-Partridge.

Arboricola intermedia. *Toung-kha*, Burmese.

The hill-partridge of Arakan is so like the Tenasserim variety of the rufous-throated that it seems ridiculous to make a "species" of it, seeing that the rufous-throated itself varies according as it inhabits the Himalayas or Tenasserim. In the Arakan variety the black of the throat is concentrated into a patch reaching up to the chin, though there are specklings elsewhere; the black border-line between the chestnut throat and grey breast is wanting as in the Tenasserim rufous-throated birds.

Hume very appositely points out that "both in this race and in the Himalayan one, specimens occur in which the black spots on the throat are large, and almost coalesce in some places; and in our present species the throat patch is at times small and dotted with pale ferruginous, showing that it is nothing but coalesced spots," and he therefore thinks it no more worthy of separation than the Tenasserim form.

Silly distinctions like these make ornithology ridiculous, and it is curious that Hume, holding the rational view above quoted as to the close alliance of this bird with the ordinary rufous-throat, wasted a plate on it.

The only point of interest on record about the bird is that its white eggs have been taken in Manipur, where, as well as in North Cachar, and in the Naga Hills, it is now known to extend. The eggs were found in May, and six was the set found.

White-cheeked or Black-throated Hill-Partridge.

Arboricola atrigularis. *Duboy*, Assamese.

The alternative English names of this species express two noticeable points in its coloration; its Hindustani name is *Peura*, which really seems to mean any sort of hill-partridge, these uninteresting species not attracting the special attention of natives. The Chittagong name is *Sanbatai.*

Besides the white cheeks and black throat, this species is notable for not having any chestnut streaks on the white-spotted flanks, the ground of which is grey like the breast. The sides of the neck are buff, speckled with black, and the black throat patch frays out into spots as it joins the breast, and just above the juncture is mixed with white. The back is barred with black as in the common hill-partridge, but the legs are light orange-red. The red skin round the eye, commonly found in these hill-partridges, is noticeable even through the face-feathers in this species.

This is a bird of our eastern hills, its western limit being Assam; it ranges east to Cachar and south to Chittagong, but does not go north of the Brahmaputra. It is fond of dense forest and bamboo-jungle, and is usually only seen singly, though in

Pl. IV.

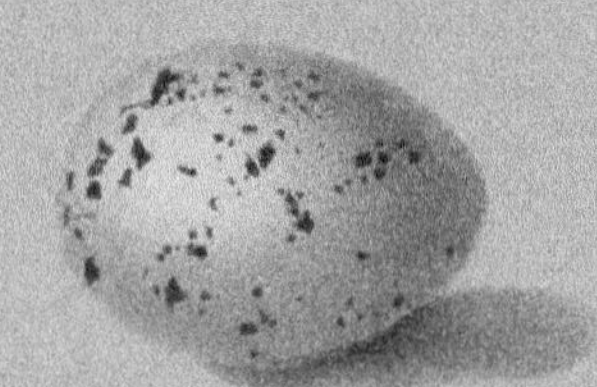

Coturnix coromandelica.

Coturnix coromandelica

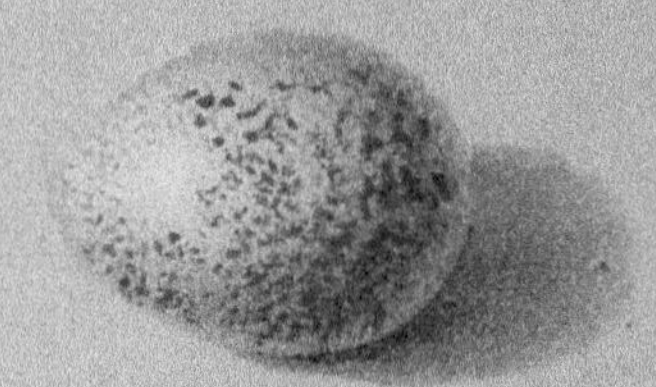

Coturnix coromandelica

Grus antigone

Fuligula nyroca

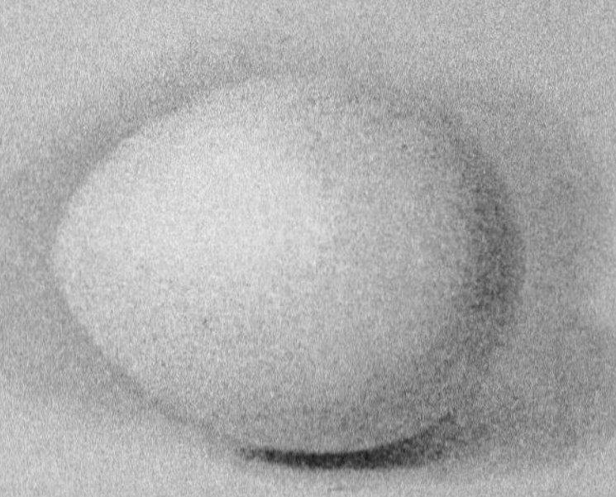

Arboricola atrogularis.

A. W. Strutt Del.

F. Waller Chromo Lith.

such cases it is suspected that often a covey may be really present, but refusing to rise simply because one of their number has been disturbed, for sometimes parties of half a dozen may be seen. Their call, according to Cripps, is "a rolling whistle, *whew, whew*, repeated many times, and winding up with a sharper and more quickly uttered *whew*. As is usually the case with these partridges, the call can be easily imitated, and by such imitation they are most readily shot; in the ordinary way, as with the common hill-partridge of the west, they afford only chance-shots, and are only worth picking up casually when such opportunities occur. They rise with a loud whirr and whistled alarm-call, and fly well and fast; but will not fly at all if they can help it. The eggs are white, and have been found at the foot of trees on shaded teelahs, of about two hundred feet high; the nest was a scrape lined with leaves and twigs, and four seems to be the full set, as they have been found incubated.

Red-breasted Hill-Partridge.

Arboricola mandellii.

This is the rarest and at the same time the most striking of our hill-partridges, so that it is curious that only very few specimens have been obtained so far, and these in Bhutan and Sikkim, always at low elevations. It is of the usual olive-brown seen in these partridges above, spotted distinctly with black; below it is dull grey for the most part, with the chestnut markings on the flanks not very distinct, and the white spots small. But the throat and breast are very distinctly and handsomely coloured, the former being bright chestnut or rusty, speckled with black at the sides, while the latter is of a much deeper shade, verging on maroon; between the two shades there is a well-marked double collar on the front of the neck, white above and black below. The hen is perhaps rather duller than the cock, but there is no certainty about this, nor about the colour of the legs, so that even the appearance of this bird is not fully known. As far as size goes, however, this species is markedly smaller than the rest of the typical hill-partridges.

It seems to frequent heavy jungle on damp ground.

Brown-breasted Hill-Partridge.

Arboricola brunneipectus.

The light buffy colour of the under-parts and face of this species make it recognizable at once among our hill-partridges; on the breast the buff is strongly tinged with brown, and on the flanks with grey; and these flank-feathers have black tips as well as white spots, but there are not the usual streaks of chestnut in this region. The throat-feathers are marked with black, and the red of the skin shows through them more or less. The brown back is barred with black, the barrings being bold and strong: and the back of the neck is black. This is one of the red-legged species—in fact, it is noticeable that most hill-partridges are red on the legs as well as round the eyes.

In spite of the noticeable distinction in colour between this and the Arakan hill-partridge, the Burmese name *Toungkha* appears to be applied to both, just as in the Himalayas *Peura* is rather a generic than a specific name.

It is in the Ruby Mines district and the eastern hills of Tenasserim and Pegu, that this hill-partridge has been found so far, living in evergreen forest at any elevation up to 4,500 feet; they especially frequent densely-wooded ravines and nullahs, and it has been noticed that the green-legged hill-partridge (*Tropicoperdix chloropus*) and the present bird are never found in the same valley, which points to some competition or conflict between them. In Pegu, where Oates observed this, it was found that these hill-partridges were only found on the eastern slopes of the hills, the western declivities being only clothed with jungle too thin to meet their requirements in the way of cover. Here they were living almost entirely on hard seeds.

At Thoungyah, Hume's collector, Darling, found them so common that he saw two or three coveys every day; from six to ten formed a covey, but they were not easy to get. They uttered a "soft cooing whistle" as they ran about scratching amongst the leaves for their food of insects, little snails, and seed, but when treed by a dog their whistle became shriller and higher until another answered. The sight of a man drove them

ARBORICOLA BRUNNEIPECTUS

scattering into the dense underbrush. Tickell, who first described the species, says they cannot be flushed more than once, and when they had settled in the shelter of a bush they would squat till they were within a yard of the muzzle of his gun. He heard them now and then emit a purring note when creeping about in the cover. Except that it is believed to breed in May, nothing is known about the nesting of this species.

Green-legged Hill-Partridge.

Tropicoperdix chloropus.

The green-legged hill partridge is easily recognized, not only by its green legs, but by the coloration of the flanks, which, instead of being grey with white spots, as is usual in the group, are light reddish brown boldly marked with black. The upper throat is white, but the lower chestnut, speckled with black as in the rufous-throated species, and the breast is really brown like the back, which may easily lead to confusion with the brown-breasted hill-partridge so-called, which often lives in the same district as, though not exactly alongside, this one, and certainly has the breast no browner; indeed, it is a particularly light, buffy-tinted bird.

The colour of the back in the present species is brown, with rather fine and close markings of black. It has a remarkable peculiarity in the plumage of the upper flanks, where there is a white downy area, overlaid by the ordinary feathers, and further hidden by the closed wing. The use of this peculiar patch is not known, and it would be worth while to study the bird in life to see if it is displayed at any time, or if, like the down patches on the breast and rump of a heron, it secretes a powdery substance.

The green-legged hill-partridge is not found in India proper, but is a Cochin Chinese and Burmese bird, occurring, like the brown-breasted species, on the Pegu hills, though not on the western faces of these, and also inhabiting Tenasserim, where indeed it was first obtained. In Pegu, Oates noted, as above observed, that it did not frequent exactly the same localities as the brown-breasted kind, even in the same district, and Tickell, who discovered it, remarked that in Tenasserim it actually avoided mountains, and frequented low-lying jungle on dry undulated ground. Davison also observed that though sometimes found in heavy forest, it preferred a thinner growth, and unlike the rufous-throated hill-partridge, would settle again on the ground when flushed by a dog, instead of perching.

In other respects, in their skulking habits, and in having a double-whistled note, these birds conform to the ordinary hill-partridge type, but the small differences of detail are interesting, as they are correlated with differences in structural points—the white down-patch under the wing, and the absence of a peculiar bony ridge over the eye which the typical hill-partridges possess.

Malayan or Charlton's Hill-Partridge.

**Tropicoperdix charltoni.*

Charlton's hill-partridge is a Malayan bird, ranging as far east as Borneo, which Hume apparently included, because it was said to have been sent from the hills of South Tenasserim, though he doubted its actual occurrence there. As a matter of fact, however, it is very difficult to draw exact lines in the distribution of skulking forest birds like partridges of this type, and South Tenasserim forms the northern limit of several Malayan species,

The present bird is closely allied to the green-legged species, and also has green legs, although Hume thought they were red and figured them accordingly; I speak from examining live specimens at the London Zoo. The two species, however, present marked distinctions in marking and colouring, the present bird having distinct orange-rusty cheek patches and a plain chestnut band beneath the black-speckled white throat. The black on the buff sides takes the form of distinct vertical bars, and that on the brown back is in very fine pencilling, finer than any found in our other hill-partridges, and recalling the delicate markings of some ducks. These two green-legged forms make a small

---

**Perdix* on plate.

PERDIX CHARLTONI

separate group of their own; the present species has the usual habits, living in coveys and skulking in forest cover, and feeding on the usual mixed diet. It is, however, not particular about elevation, going down to the bases of the hills, though not found off them. The note is described as a double whistle.

Red-crested Partridge.

Rollulus roulroul. *See-oul,* Malay.

To call this bird a wood-*quail,* as Hume does, is distinctly misleading, for there is nothing specially quail-like about it, since it closely resembles the hill-partridges in size and shape, being round and dumpy, with very short tucked-in tail, and, nevertheless, high on the legs. It has not, however, the long claws of the wood-partridges, and the cock would look among them like a rajah among coolies, with his bushy crimson aigrette and suit of deep-blue velvet. Before the crest springs a bunch of bristles, and this and the red eye-ring and legs are common to both sexes; otherwise the cock and hen could hardly be imagined to belong to the same species, since the lady, though devoid of a crest, is almost as brilliant as her mate, being clad in leaf-green, a most extraordinary colour for any partridge, or for the hen of any game-bird for that matter. Her wings are chestnut, whereas the male's are only brown.

The chicks are of a brown colour both in the down and in the first plumage, as has been ascertained from specimens bred in England.

The red-crested partridge is a bird of the Malay region, extending from South Tenasserim down through the Peninsula and the islands to Java; in Sumatra it is called *Banisel.* In spite of its very aristocratic appearance—there is something about it that always reminds one of a pigmy peafowl, in spite of the short tail—it has much the same habits as the common-looking hill-partridges, associating in small parties which frequent heavy forest, and having a whistling note, described as mellow and pleasant. The red-crests differ, however, in the detail of not being nearly such inveterate scratchers as the hill-partridges, and in quickness and alertness are compared to quails, like which they freely run about.

Both cocks and hens are found in the coveys, which number up to a dozen. The cocks have no spurs, and nothing seems to be recorded about their ways in nature when breeding, though it is probable they break up into pairs at this time, and some fighting may take place. In captivity they are not at all shy.

The cock is, at any rate, devoted to his hen, and, as I have seen myself in captive specimens, will feed her with tit-bits after the manner of the common farmyard fowl. The egg is known to be buff in colour.

The food consists of seeds, berries, tender leaves, &c., and insects; as to the quality of the birds themselves for table nothing seems recorded.

Ferruginous or Chestnut Wood-Partridge.

Caloperdix oculea. *Burong trung*, Malay.

This pretty bird approaches the typical partridges in shape more than the hill-partridges do, and its bright, chestnut plumage is very distinctive and marks it out from any other found with us; on the sides and lower back the chestnut is diversified by black markings, which in the latter region indeed obscure the red; the upper back is variegated with black and white. The only sex difference is the presence of spurs on the legs of the male; the legs themselves are green. As in so many of our game-birds, more than one spur may be present.

Like the red-crested wood-partridge this is a Malayan bird, gaining a place in the Indian list by penetrating into South Tenasserim; but it does not range further east than Sumatra in the other direction. It is a lover of heavy jungle, where it feeds on berries, seeds, and insects; but the places where Mr. Hume's collectors found it were so lonely that there were no inhabitants, and no paths except those made by the local big game; and they never even saw the birds they got until they were caught. Even the Malays knew nothing about it, so that this, one of our handsomest small game-birds, remains eminently

ROLLULUS ROULROUL

a subject for research. A hen bird weighed only eight ounces, but males would probably be a little heavier; but the bird is not a big one, measuring well under a foot.

Jungle Bush-Quail.

*Perdicula asiatica.** *Lowa*, Hindustani.

This funny, thick-set, stout-billed little bird, though not so big as the ordinary quail, is really a pigmy relative of the ordinary grey partridge, which it much resembles in many of its ways, even to the detail of getting into a temper when blown upon in a cage. Like larger partridges, it has an easily distinguishable, though small, tail, and even a rudimentary spur on the legs of the cock. The bright chestnut head will distinguish it from all our other quails, except its near relative next to be mentioned; both hen and cock have this ginger headpiece, but the rest of the hen's plumage is a simple light brown, while the cock sports a zebra waistcoat of black and white. Young birds have the plumage brown with pale streaks, and no reddish tint on the head. The bill is black, and the legs orange.

The native name *Lowa*, applied to this bird, is also given to button-quails, and evidently means some bird which is much like a quail, but not the exact thing. Other names—*Juhar*, used in Manbhum, and the Canarese *Kari Lowga*, Sonthal *Auriconnai*, and Telugu *Girzapitta*, bearing witness to the marked personality of this little fellow, which is a favourite fighting bird with natives, combating with more noise and fury than even the grey partridge.

It is confined to our Empire, and in that to the Peninsula and North Ceylon; while even in this restricted range it is local, although a dry or moist climate does not affect it much; nor is it particular about elevation, ranging up to 5,000 feet. But it wants its location dry underfoot, and frequents wooded and broken or sloping ground, though it will come into grass and stubble to feed, and is quite contented with scrub cover; but cover of some

---

**cambaiensis* on plate.

sort it must have. Although a ground bird, it will take to trees if put up by dogs, like the grey partridge, and it also has the partridge habit of sociability carried to an extreme, for, though sometimes found in pairs in the breeding-season, it is usually found in coveys, even up to a score in number, which pack very closely, and forage about together like a flock of guinea-fowls in miniature.

This extreme sociability, which, as in the great snow-cocks, extends so far that young ones may be seen in company with several of their elders, makes it strange that the birds should be so pugnacious, but probably the ties of friendship only hold for the same covey, which are mostly, no doubt, near relatives. Strangers are probably barred by flocking birds as well as solitary ones; and in the case of another well-known social bird, the rat-bird or common babbler (*Argya caudata*), two flocks working the same hedge have been seen to meet and fight with such fury that they adjourned to the road to fight out the matter in couples. Be that as it may, this bush-quail is commonly captured by means of a decoy-bird in a cage set with nooses, like the grey partridge; for more sporting methods of capture it is not of much use, despite a remarkably tame disposition, for when pressed the whole covey explodes, as it were, in all directions, whistling and whirring—including sometimes, as Tickell says, a close shave of the sportsman's countenance—and each member drops as suddenly as it rose after just shaving the bushes in a very swift flight of a couple of dozen yards, rapidly reassembling to the peculiar trilling pipe of the head of the covey. When bagged bush-quail are not much to boast of, weighing little over two ounces, and being very dry. They feed chiefly on seeds of grass and millet, and are pretty certain to be found in ragi stubble; insects are also often consumed.

They breed very late in the year, beginning in September, and eggs may be taken in February; the nest is under a tuft of grass or a bush, and fairly neatly made, and the eggs pale creamy and as few as four or as many as seven in number.

CALOPERDIX OCCULEA

Rock Bush-Quail.

*Perdicula argunda.** *Sinkadeh*, Tamil.

This, the only near relative of the last species, has been very much mixed up with it; the native name in Hindustani is the same, though in Canarese it is distinguished as *Kemp Lowga*, and in Telugu as *Lawunka*. The wonder is that the birds themselves have not got mixed up, especially as they sometimes occur together; if they inhabited separate areas altogether one would be inclined to regard the present one as only a local race of the other. In Hume's plate, as he points out, a female of the jungle species does duty in the foreground (the standing bird) as a representative of this one, and the plate is lettered for that species. Yet there is a positive difference; in the present species both webs of the pinion-quills are marked with buff, while such markings are only on the outer web in the other. The differences elsewhere are mostly comparative, as is usual in local races rather than true species; the rock bush-quail is larger, has the red of the head much duller, and, in the cock, the barring of the under-parts broader. The hen has the top of the throat whitish, and a whitish abdomen, but both lack the eyebrow-stripe of white found in the other kind. The rock bush-quail is not found in Ceylon at all, and, though a Peninsular bird, is on the whole differently distributed, since it is usually found on different ground, the two species largely replacing each other; though, as has been said above, they may at times be found together.

The rock bush-quail is not so fond of cultivation or elevated land, being more a bird of dry sandy plains or hillocks, where the only vegetation is scanty scrub; it is, in fact, a bird of the open wastes, though, as its name implies, it especially likes rocky ground. Except for this choice of location, it is very like its ally in all its ways; occasionally takes to trees when disturbed, and goes in the same coveys, which go off in the same sort of feathered *feu de joie* when flushed. It affords fair sport when worked with dogs among the low scrub, but is as dry eating as its relative; though, as Hume rather sarcastically admits, it will

---

**asiatica* on plate.

make a good pie in conjunction with a whole larder-full of accessories which he enumerates. It is also captured by natives as a fighting bird, but is not so much used or so highly esteemed.

Like the jungle bush-quail, it is sociable even in the breeding-season, which lasts from August to March, young birds being seen in company with several old ones; the eggs are creamy white, and four or five in number. Hume says he has noticed no difference in the note, which makes it the more remarkable that the birds maintain their distinction, since a difference in language is often found between otherwise similar birds, notably the common and grey quails.

Painted Bush-Quail.

**Microperdix erythrorhynchus.* *Kadai*, Tamil.

The painted bush-quail is distinguished from all our other quail-like birds, except the next, which is merely a local race of it, by its red bill and legs and the large black spots on its brown plumage; the cock is distinguished from the hen by having a white throat and eyebrows, and black face, chin and cap, the head in the hen being, like the belly in both sexes, of a chestnut colour.

In young cocks the black cap is the first sex-mark to appear. Although approaching the thick-billed bush-quails rather than the typical quails, this bird has a smaller beak, no indications of spurs, and rises with less of a whirr.

It is found on the hills of Southern India, the Nilgiris, Pulneys, and Shevaroys, and ranges up the Western Ghauts as far as Bombay. It is also to be found in hilly tracts east of these Ghauts, and has strayed even to Poona. In habits it is a bush-quail, not a typical quail; it frequents the outskirts of jungle and rocky ground interspersed with low cover, and associates in coveys, the members of which generally all go off together, but in different directions, when flushed. Sometimes, however, they rise independently and at intervals of several minutes; unless

* *Perdicula* on plate.

PERDICULA CAMBAIENSIS

hunted out by a dog, however, they very strongly object to rising twice.

Their call, according to Davison, is "a series of whistling notes, commencing very soft and low, and ending high and rather shrill, the first part of the call being composed of single and the latter of double notes, sounding sometimes like *tu-tu-tu-tu-tutu-tutu-tutu*, &c." By the use of this call, given low and cautiously at first, the scattered covey is reunited again; the call seems to have something of the ventriloquial character.

They resemble the thick-billed bush-quails in being very quarrelsome in spite of their sociability, so that they are readily captured in a trap-cage with a call-bird in the inner compartment. The ferocity of some of these harmless-looking little game-birds, and their powers of hurting each other, are indeed remarkable. I remember once seeing one of this species put into a cage where there were others, and after being left unwatched for only a few minutes, it had to be taken out and killed owing to the cruel mangling its head had undergone at the beaks of its new associates—this again exemplifies what I suggested in the case of the jungle bush-quail, that charity begins (and ends) at home with these birds.

The painted bush-quail is of a tame nature, and likes to live near cultivation and roads, where grain can be obtained; it especially likes millet, but also, of course, feeds on wild small seeds, which, with insects, form its main diet. It runs so swiftly, says Miss Cockburn, as to look like a little brown ball rolled along the ground. No one seems to have made any special notes about the table qualities of this bush-quail; its weight is from about two and a half ounces. It is resident in its chosen haunts, and, except perhaps in May, June and July, eggs may be found in any month of the year in one place or another; on the Nilgiris Miss Cockburn found that the birds bred twice yearly; in the first quarter of the year and again in autumn. As is so usual with this group of birds, the nest may be a mere scrape in the soil or have a lining of grass; the cream-coloured eggs, which are ten or even more in number, are described by Hume as intermediate in size and colour between those of the grey partridge and the rock bush-quail. The young are exceedingly active, and start on the move a few minutes after being hatched. Their down is dark with three longitudinal cream-coloured stripes, and both parents accompany them—indeed, one very unsportsmanlike way of capturing old birds is to dig a hole, catch some of the little innocents, and put them in, when the parents soon jump down to them, and a cloth is thrown over the lot.

Eastern or Blewitt's Painted Bush-Quail.

**Microperdix blewitti.* *Sirsi lowa*, Hindustani.

This race of red-legged or painted bush-quail found in the Eastern Central Provinces, where the ordinary kind does not occur, differs from the type in no important particular, and here again I wonder Hume wasted a plate on it; the result was not happy, as the legs and bills of the birds are there represented as yellow, whereas they ought to be red—a mistake of the artist's, of course. The distinguishing points, as usual in a local race as compared to a true species, are merely comparative; a smaller beak, greyer tone of plumage, dull pinkish tinge over the abdomen, instead of only on the breast and along the flanks as in the typical form, greater extension of the white on the head of the cock at the expense of the black face and crown, and finally smaller size, which barely reaches two and a quarter ounces, while in the other race it runs from this to over three. It was said by Blewitt, the sender of the first specimens to Hume, to be delicate and well-flavoured.

He also found that it went in coveys of sometimes more than a dozen, living in forest, grass, and scrub on hilly ground. He noted, from native information, the breeding-season as November to January, but Thompson gives it as June and July, soon after the rains begin, the young flying in September. The fact probably is that this race and the typical one both breed at any time which local conditions make convenient for them. Thompson's note, however, that the male in the courting-time often repeats a loud single note, is worth quoting, and also his

* *Perdicula* on plate.

PERDICULA ASIATICA.

experience in finding these birds frequenting " long grass on the banks of nullas and rivers." Blewitt gave the notes as "more soft and melodious " than that of the others, by which " others" presumably he meant the thick-billed bush-quails ; so this need not indicate a difference in voice between the two races.

Hume's Bush-Quail.

Microperdix manipurensis. *Lanz soibol,* Manipuri.

This bush-quail can really be called a grey quail, since its prevailing colour above is slate, with no tinge of brown, but diversified by black markings ; the under-surface is mottled with buff and black, the buff predominating as large spots, almost concealing the black groundwork. The legs are orange, and the only difference between the cock and the hen is the dark reddish-chocolate face of the former sex.

In Manipur this bird is fairly common, but very hard to get, or even to see, as it haunts high grass, and even after this is burnt is still difficult to discover, owing to its dark colour harmonizing with the burnt stubble. It affects the neighbourhood of water, and keeps in coveys which run closely packed.

Eggs have been obtained in Manipur, but not preserved; they are marked with blotches of brown and black on a greenish ground. Our knowledge of this quail is due entirely to Hume, its discoverer, who got a few specimens with a great deal of trouble in beating and cutting down huge quantities of the high grassy cover wherein he observed them, and to Captain Wood, who shot numbers of them as lately as 1899.

Since this Mr. C. M. Inglis has procured specimens of a very nearly allied species, named by Mr. Ogilvie Grant after him *Microperdix inglisi,* in Goalpara ; the differences from the Manipur bird are very slight, the comparative scantiness of the black markings being the most noticeable. This form is also suspected of occurring in the Bhutan Duars. Its Goalpara name is *Kala goondri.*

Common Quail.

Coturnix communis. *Bater*, Hindustani.

When Marco Polo had travelled in India, he ventured on the observation that all the birds and beasts there were different from those of Europe, except the quail; and though, as we all know, this is by no means correct, it is nevertheless striking to find a bird so well known in Europe as the quail also equally familiar in the East. There are, however, several little game-birds which go by the name of quail in India, and it is as well to point out the distinctions of this species, which is *the* original quail, from all the rest. In the first place, it is distinguished from most of them by having such a very indistinct tail, the real tail-feathers being so soft and so exactly like those of the rest of the hind-parts, that it is difficult to sort them out, as it were; this character is also found in the button-quails, which are not true quails at all, but these only have three toes instead of the usual four. The soft tail and four-toed feet will, then, distinguish the common quail from all familiar quail-like birds except its allies the rain and painted quails, and it is larger than either of these, to say nothing of other differences; the closed wing measures at least four inches, whereas it does not attain this length even in the rain-quail, while the painted quail is far smaller again than this. The distinctions of the rare Japanese quail will appear later.

There is nothing very noteworthy in the general plumage of the common quail; it is often called grey quail, but the name is misleading, as the plumage is not grey or even greyish, but light brown, well variegated with black, and diversified above by well-marked longitudinal streaks of cream-colour. The pinion-quills of the wing are drab, barred with buff, and this is the chiet distinction from the rain-quail, in which these quills are uniform drab with no markings.

The difference between the sexes is not apparent on the upper plumage, but is noticeable enough below, where the cock is a plain clear uniform buff, with the throat marked with sooty-black on a whitish or brownish-red ground. In the hen the

PERDICULA ERYTHROHYNCHA

throat is always all white, but the breast is marked with short blackish streaks as in a lark, and the general tone below is paler and not so buffy, more of a cream-colour. The largest quail of this species are hens, but many cocks are as big as most of their mates ; the weight ranges from 3·2 ounces to 4·62 ounces—a big variation for so small a bird, but a good deal has to be allowed for condition, the quail being a bird which under favourable circumstances gets very fat. Nothing need be said about its value for the table, since it has been esteemed in this capacity for untold ages, and therefore persecuted by man longer and more thoroughly than any other species of bird whatever. Nevertheless, it is still exceedingly common almost throughout the north temperate parts of the Old World, and in India, which is one of its great wintering-places, is the most abundant of all game birds during the winter months, though its numbers vary much in different years, and also the wideness of its distribution. A few—a very, very few—remain to breed here, but nearly all normally leave us by the end of April.

It is, indeed, essentially a long-distance migrant, the only one of its family; indeed, most of them, whether pheasants, partridges or other quails, are considered good fliers of their kind if they go ordinarily a mile without alighting, while this little quail crosses both the Mediterranean and the Red Sea, though often absolutely worn out by a long passage. There is a great loss of life during migration owing to the powers of flight of the birds being barely sufficient for such long journeys, and evidently thousands of years of evolution have failed to completely adapt this bird to habits so unlike those of its kin. The difference in flying power would never be appreciated by observers of the ordinary habits of the quail, for when flushed in the fields it seldom flies a quarter of a mile, or rises more than a yard or two from the ground; its flight is very straight and steady, and performed by a continuous quick beat of the wings. Although swift, it is not a difficult bird to shoot, and where it is common may be shot in enormous numbers ; bags of a hundred brace in a day are mentioned by Hume, and yet the birds are not at all gregarious, but get up and fly singly, though when on the move they do travel *en masse*. When migrating they travel at night, and must often go very high, as they cross the Himalayas on one of their migration routes; in fact, the bulk of our quail in India come to us in this way, arriving usually during the first half of September. These have summered in Central Asia; but a further set come in on our north-western coasts, from Arabia and Africa, and these arrive before August is out.

Once arrived in India, the quail proceed to distribute themselves according to circumstances; a place may be swarming with them one day and deserted the next, for they still keep moving on in many cases. They never reach the extreme southern and western parts of our area, however, not penetrating as far as Ceylon, nor have they been found in Tenasserim, while even in Chittagong and Burma they are rare. In Lower Bengal, also, they cannot usually be very common, for I only heard of them as abundant in the winter of 1900-1901, out of seven I spent in Calcutta. In years of scarcity they are common in Central and Southern India, and the worse the season the further they naturally go; but normally Upper India is their stronghold, though they are only really abundant locally. In March they are commonest, because then they are drawing up for the northward migration. The variation in their visiting numbers is estimated by Hume as probably one of many millions, and this is quite likely, for this quail is probably one of the most numerous birds in the world. Man, it is true, is a great enemy, as has been said; but, on the other hand, he creates conditions favourable for the bird, which is quite at home in cultivation, and only avoids deserts, swamps, and forest, which are just the sort of country which man desires to see converted into cultivable land, and which without him form the major part of the earth's surface. The common quail not only finds shelter, but food in human cultivations; for although in the wilds its food must consist only of grass-seed, small berries, and insects, it gladly feeds on grain, especially the various kinds of millets. When the crops are reaped, it takes to bush-jungle and a diet of wild produce. Quail feed in the morning and evening, and probably also at night, for in captivity they are active then; by day they are very sluggish, and may even be trodden upon sometimes before rising, though at others they

$\frac{3}{4}$

PERDICULA BLEWITTI

will run some distance. When winged, they are easily lost, as they hide adroitly, and will readily "go to ground" in any hole.

When at ease their note is a low whistling chirp, but is harsher when they are forced to rise, and the male's spring call is very distinct, a loud clear trisyllable, of which many renderings exist. Mr. E. Kay Robinson's "Dick, be quick" expresses it best to my ear. Although possessed of but a small bill and devoid of spurs, the cock is intensely quarrelsome, and quail-fighting is as popular a sport in India now as it used to be in ancient Greece. This quail is a very prolific bird, laying as many as fourteen eggs, but such as breed in India do not appear to lay over ten; the nest is made of a little grass, of course on the ground, for, like all typical quails, this species never even perches. The eggs are very distinctive in appearance; they are large considering the number laid, measuring more than an inch in the long diameter, and are marked, generally heavily, with chocolate on cream colour. Such quail as breed here lay in March and April, but these are, no doubt, usually "pricked" birds, though in 1872 these quail bred freely about Nowshera, probably influenced by the backward season of that year; but there appears no general tendency in Asiatic common quail to become residents, as they often do in some other countries, notably Spain and Ireland. Probably the competition of other birds of similar type is the deterrent to their colonization of India, for the strong point of the species appears to be the power of flight which enables it to occupy ground which other birds of the family can never reach.

The many names of this bird probably mean "quail" in general in most cases—*Butairo* in Sind and *Batri* in Bengali recall the Hindustani name; *Bur-ganja* and *Gur-ganj* are used at Poona, and *Burli* at Belgaum, while in Tamil and Canarese the names are *Peria-ka-deh* and *Sipale haki*, the Telugu name being *Gogari-yellachi*. *Botah Surrai* is the Assamese name, and *Soibol* in Manipur, while the Uriyas use *Gundri*. When mentioned specifically the species is distinguished in Hindustani as *Gagus* or *Burra Bater*.

Japanese Quail.

Coturnix japonica. *Udzura*, Japanese.

The male Japanese quail is easily distinguished from the male common or grey quail, to which it is very closely allied, by having a brick-red throat with no dark marking or only a central streak. In the hens the difference is chiefly to be found in the structure, not the colour, of the feathers of this part, the Japanese bird having the throat-feathers noticeably long and pointed at the tips; at the sides of the throat these elongated feathers are edged with reddish. In both sexes the reddish tint on the flanks is brighter than that found in the same place on the common quail.

Hens of this species have been shot in Bhutan, Karennee, and Manipur, and as this is the ordinary quail of the mainland of far eastern Asia, as well as Japan, it is quite probably a common winter visitor to the eastern parts of our Empire. The only noteworthy difference in the ways of this bird and the common quail appears to consist in the note of the male, which is said to be deep and hollow, of several rapidly-uttered syllables.

Rain-Quail.

Coturnix coromandelica. *Chota bater*, Hindustani.

Although the cock rain-quail is noticeably distinguished by the black streaks—in old birds coalescing into a black patch—on his more warmly-tinted breast, and by the purity and distinctness of the white and black of his throat, the species is very commonly confounded with the common quail, and it must be admitted that the hens are almost exactly alike. In this bird, however, there is none of the light chequering on the pinion-quills in either sex, and it is smaller altogether than the common quail, not exceeding two ounces in weight.

When used to it one can always pick out even the hens, without looking at the quills, by their brighter colouring and smaller size, which is conspicuous enough to distinguish this bird even in flight. The same native names, however, usually

COTURNIX COMMUNIS

are given to both, though the Telugu speakers call this bird *Chinna yellichi*, and the Tamils *Kade*, while *Chanac* is used in Nepal. This species, although to a great extent locally migratory, is a purely Indian and Burmese bird; but it does not extend to the confines of our Empire, being absent from Kashmir in one direction and Tenasserim in the other, while it is not found in Ceylon, attempts to introduce it (and the common quail also) having apparently failed.

Its name, rain-quail, has reference to its appearance in certain districts coinciding with the opening of the rains; these are the drier parts of Upper India and Burma, and it visits these to escape, apparently, from the damp in the more low-lying and rainy tracts, where, as in Lower Bengal, it is common enough in the dry months. It is generally a bird of the plains, but on the advent of the rains will penetrate up to 6,000 feet in the Himalayas and Nilgiris. In the Deccan it is resident, and also in parts of Southern India.

It frequents the same sort of low cover as suits the larger quails, and the two may often be flushed in the same locality; but although it comes freely into grassy compounds, it is not quite so much addicted to cultivation, preferring wild grass-seed to grain. It also feeds on insects, and Hume records having found one which had fed on the scarlet velvety mite, a remarkable article of diet, as he says, usually avoided by birds.

On the whole, however, there is nothing noteworthy in its ways to distinguish it from the common quail, except its very distinct two-syllabled note; it is just as pugnacious, is kept for fighting, and fattened for food in the same way. Hume thought, however, it was slightly inferior as a table bird. Like the common quail it is found in pairs or singly, not in coveys. Its breeding season lasts about half the year, from April onwards, sometimes even to November, and the eggs are in some cases very like those of the common quail, but they are smaller, and vary enormously, some being finely peppered and freckled, and some marbled. The ground colour also varies from a decided buff to nearly white, and the markings may be blackish, olive, or purplish brown. But only one shade is found on one egg, and all the eggs in a set, which does not exceed nine, unless two hens lay in one nest, as often happens, are usually much of the same type. The cock, which feeds the hen during courtship, keeps close at hand during incubation. The very scanty nest is placed among crops or moderately high grass. The main breeding-ground of this species in India is in the Deccan, Guzerat, and Central India; it appears to be much persecuted by vermin, for where the birds are breeding freely an enormous number of broken-up nests are to be found.

Painted Quail.

Excalfactoria chinensis. *Khair-butai,* Nepalese.

Only about the size of a sparrow, the painted quail is not likely to be mistaken for any of our game birds, except perhaps the even tinier little button-quail, from which the darker colour will distinguish it on the wing, and the yellow, four-toed feet in the hand; it is quite a sporting bird, too, and when flushed flies for fifty yards or more, low over the grass. Close at hand, a very striking difference is observable between cock and hen, the former having a blue-grey breast and sides, and the centre of the under-parts rich chestnut, while those handsome colours are well set off by the characteristic black anchor-mark on a white ground on the throat of the true quails, the colours in this species being as distinct as in the rain-quail. The young cock is at first brown below like the hen, but gets his full plumage in little over a month. It is only when a pair have fledged young that these quail are seen in coveys, otherwise they are found singly or in pairs.

The cock is much attached to his mate, and feeds her with insects; besides the chirping alarm-note when flushed, he has a distinct trisyllabled call, *tee-wee-wee.*

Like the rain-quail, this bird does not leave the Indian Empire (although it has a wide range outside it to the south-east, even to China) but is locally migratory within it. It is found at one time or another almost all over India and Burma, and in Ceylon, but it is essentially a bird of moist

COTURNIX COROMANDELICA

districts, and absent from the dry regions of the north-west. It ascends the Himalayas into the temperate region, and in Lower Bengal, where so many widely distributed birds are wanting, it is quite common in the cold weather. To the foot-hills of the Himalayas and districts adjacent it is a rainy-season visitant, and immense numbers arrive in Pegu at the beginning of May. The favourite haunts of these tiny birds are open, moist grass-land, and they frequent the grassed lands of paddy fields, the paddy-stubble itself, and scrub-jungle. They feed on grass-seeds and insects, but will also take millet. Where bigger quail are scarce they may be found worth shooting, if anyone cares to expend powder on birds which do not weigh at most more than a couple of ounces; I never heard of anyone eating them.

Judging from their habits in captivity, they are to a considerable extent nocturnal; but they may be seen feeding outside of the grass in the early morning, and are not very shy, though when once flushed they much object to showing themselves again. They nest in Ceylon and in the Malay Peninsula as early as March, but further north in June and later, up to even the middle of August in the Sub-Himalayan tracts. The nest is the scanty affair one expects from a quail, and is placed among grass, containing about six eggs, buff or pale drab generally, somewhat peppered with brown. Considering the size of the producer, they are remarkably large, many being an inch in the larger diameter; the incubation period is three weeks, and the minute chicks are dark with pale streaks.

In Ceylon this pigmy quail is known as *Pandura* or *Wenella-watuwa*, and as *Gobal-butai* in Oudh. *Kaneli* is also a Nepalese name.

Bustard-Quail.

Turnix pugnax. *Gundlu*, Hindustani.

This quaint little bird, which may be easily taken at first sight for a quail, is yet at once distinguishable from our true quails by having no hind-toe, which applies to all the members of its family found with us. The said family is quite a distinct one, and the birds composing it are often called in books hemipodes—

a silly name, because it means "half-foot" and only one toe out of the normal four is missing. Nevertheless, it is better than button-quail or bustard-quail, because the birds are neither quails nor bustards. The present species is the most widely distributed in our limits, being only absent from elevations over 7,000 feet in the Himalayas, and from parts of the north-west; for, though found in Cutch and Rajputana, it does not occur in Sind and the Punjab. It extends across the rest of Asia to Formosa, including the Malay Islands. Hume figures the Eastern race as distinct, but it is not now so considered; it is merely larger and of a less reddish brown.

It may be distinguished from the other and less widely distributed hemipodes by its bluish-grey beak and legs, which mark it off from the yellow-legged species, and by being barred with black on the breast, which distinguishes it from the little button-quail, which also has a blue beak and frequently blue legs also. The hen is larger than the cock, as in all this family, and also more strikingly coloured, having a black patch running down the throat and breast.

This little bird has the general habits of the true quails, being found among grass and bush-cover, and avoiding high forest and arid tracts; it also feeds on seeds, herbage, and insects. I suspect it is more insectivorous than quails proper, its larger bill enabling it to manage insects of bigger size; it appears to care little for grain. Hume thinks that these hemipodes do not drink, but I have seen them do so in captivity, and the fact that they are not to be seen drinking when wild probably only means that they quench their thirst with dew instead of resorting to bodies of water for drinking purposes. Their more insectivorous habits—if I am right about these—would also probably imply greater independence of water, for it is animal-feeding birds which can generally best dispense with this, though among beasts the reverse holds good.

These birds are generally solitary or at most in pairs, except when a brood of young is about; they lie very close and fly only for a few yards at a time, after which they are almost impossible to raise again, and it takes a smart dog to get them up at all. Nevertheless, they migrate a little, but only according to circum-

EXCALFACTORIA SINENSIS

stances, to avoid cold in the hills or floods in the plains. Their disposition is quite different from that of the true quails, as they are singularly tame in captivity, instead of wildly nervous like nearly all true game birds; and probably quails are serious enemies to them, as I have found that hemipodes of any sort, taken out of a dealer's crate of quails, are generally much plucked, just as the tiny blue-breasted quail is. This may perhaps be the reason why this bird frequents gardens so much.

But the most remarkable point about this bird and its kin is the peculiar reversal of their sex relations. The hen, as we have seen, is the larger and finer bird; she is also the fighter, and is constantly captured by the natives as a fighting-bird, the attraction being another female in a cage, while males are never so caught. So well is the distinction known that the two sexes have different names in more than one language, the cock in Telugu being *Koladu*, and the hen *Pured*, while in Tamil he is *Ankadeh*, and the hen *Kurung kadeh*. In the Malay countries, too, the name of the bird, *Pee-yoo*, is applied in contempt to a hen-pecked man, for the cock bustard quail not only does not fight, but makes the nest and sits on the eggs. The nest varies from a mere "scrape" to a proper though loose structure made of dry grass, and often domed over. It is commonly found in the Darjeeling tea gardens in May and June, but in the plains the breeding season is later, and extends to September in Burma. As, however, eggs have been taken in March at the south end of the Malay peninsula, the birds may breed here and there almost all the year.

Only four eggs are laid, at any rate as a rule; they are short, and may show a tendency to the "peg-top" shape; they are glossy and minutely peppered all over on a dirty-white ground, and generally blotched with larger markings as well; they are about an inch long—*i.e.*, large for the size of the bird—which is much smaller than a common quail.

The note of the bird, chiefly given out by the hen, is a purring sound according to Mr. Seth Smith, who has studied the species in captivity, but Hutton says, speaking of it in the Dun, that it has a pleasing, ringing note; he also says it is brought in large numbers for sale, but this was not the case in Calcutta in my time, though the species is supposed to be common about there. It is a much nicer aviary bird—like all hemipodes—than the true quails, but of little interest to the sportsman, being scantily distributed and giving a very poor shot for a good deal of trouble.

Besides the native names above mentioned, this bird is called *Durwa* at Ratnagiri, *Karé-haki* in Canarese, and *Timok* by the Lepchas.

Indian Yellow-legged Button-Quail.

*Turnix tanki.** *Pedda daba-gundlu*, Telugu.

The yellow-legged button-quail is easily distinguished from the bustard-quail by its yellow legs and bill, and, of course, from such of the true quails as are yellow-legged, by the absence of the hind toe. It agrees with the bustard-quail in the difference of size in the sexes and in the female being more richly coloured; but the decoration is quite different, the female having a chestnut collar instead of a black cravat, and this is not permanent, being only assumed during the breeding season. The back is less variegated in this species than the last, though young birds have more marking than adults, but the most conspicuous difference, besides the yellow bill and feet, is the absence of any bars on the breast. There is practically no difference in size between this particular yellow-legged race and the blue-legged hemipode.

Although more numerous in India proper than the bustard-quail, and found in the North-west districts, where the other is absent, the yellow-legged bird does not go so high up in the Himalayas, my record of one caught by Mr. Goldstein at Darjeeling, in my book on " The Game Birds of India and Asia," being quite an exception, the usual limit of this bird's vertical range being 4,000 feet. As this was caught at night at light, it looks as if the bird were migrating, but it might have been a mere stray. This button-quail does not occur in Ceylon, and

* *jondera* on plate.

TURNIX JONDERA

its eastern limit is the Naga Hills; in Assam begins the range of the large Burmese race of this yellow-legged type.

There is little to be said about the habits of this bird, which are much like those of the bustard-quail, but it affects drier localities, and does not come quite so much into cultivation on the whole; moderately high grass is a pretty good place in which to look for it, and it is also found in grassy patches in forest clearings. Its flight is feebler and less whirring and noisy than the bustard-quail's, and it goes for even a shorter distance when flushed, dropping so quickly as scarcely to allow time for a shot, and lying so close afterwards that smart dogs may often pick it up. In captivity it shows an even tamer disposition than the blue-legged bird. The first pair the Zoo in London ever had, presented by Mr. E. W. Harper, were so tame that I have poked my finger through and touched them as they sat at the side of the aviary. This bird lays four eggs, peppered and blotched like those of the bustard-quail, in a domed nest of grass.

Mr. D. Seth Smith, now Bird Curator at the Zoo, has given, in the *Avicultural Magazine* for 1902-03, some very interesting details of the habits of this species as observed by him in the private aviary he then had. He successfully bred the birds, this being the first instance of any hemipode being bred in Britain; and found out about the seasonal change in the female's collar, and also that she gave any mealworms given her to her mate, thus showing that the moral reversal of the sexes in the hemipodes results in the hen being generous as well as quarrelsome. She did not, however, feed the chicks, and the male did everything for them as well as the sitting, which only lasted twelve days—a remarkably short period, for even a canary takes fourteen. In the aviary, which had a grassed outdoor enclosure, he noticed that the birds did not seem so much at home in the long grass itself as the painted quails, which made little tunnels in it and bolted down them, but preferred sandy spots with grass tufts here and there; this is rather at variance with Indian experience of it as a grass bird, but Tickell says it is found, in Bengal at any rate, "in open, sandy, bushy places." The young were mottled rather than distinctly striped like the young of the true quails, and were very insectivorous, refusing at first all kinds of artificial food, which the young of the true game birds nevertheless eat readily. The note of the hen is "a soft booming sound, which is more or less ventriloquial"; the male seldom calls, if at all, and all the bird utters when flushed in the wild state is a faint low double chirp. Tickell says this bird is most delicious eating, but Hume condemns it; probably both are right, the difference depending on food.

Burmese Yellow-legged Button-Quail.

*Turnix blanfordi.**

In Burma yellow-legged button-quails occur as in India, and the sportsman who pays attention to these little birds may notice that they are larger than the Indian birds of this type, if he has had opportunities for comparison. Young birds of both kinds have the same plumage, which is variable in details in both, but this eastern form does not lose the dark markings on the upper-parts with age to the same extent as the Indian specimens, which incline decidedly to a uniform drab above.

Although classed as a species, the distinction is very trifling, and of no interest to anyone except those naturalists who like niggling over local races; there seems to be nothing special on record about the habits of the variety, which are not likely to differ in any important particular from those of the Indian bird.

The race, such as it is, extends into Assam and Chittagong, while in the other direction it is found in China and even in Eastern Siberia.

Nicobar Yellow-legged Button-Quail.

Turnix albiventris.

In the grassy parts of the Nicobars and Andamans is found another button-quail of the yellow-legged type, coming still nearer to the Indian typical form, and only distinguishable by its generally darker colour and more abundant markings

* *maculosa* on plate.

TURNIX MACULOSUS

above; even this distinction can only be properly appreciated in adults. As the Andamans are mostly under forest, there are few places in which this grass-haunting bird can live, and so it is little known there.

Little Button-Quail:

Turnix dussumieri. *Dabki*, Hindustani.

This funny little midget, about the size of a sparrow, bears the same relation in size to the other hemipodes that the jack snipe does to the other snipe, and curiously enough is distinguished in two other similar ways, in having a pointed tail and brighter-coloured plumage; there is but little black in the upper plumage, and a good deal of straw-colour and bright chestnut, and the under-surface is pale and clear. The buff breast is plain in the centre, but along the sides of it are some round markings of black. The legs are usually white, but sometimes blue-grey like the bill. The characteristic superiority in size of the hen is not so striking in this species, and she has no distinctive decoration; but the young are duller and more uniformly brown than the old birds.

The little button-quail, which, I take it, is *the* button-quail, from the small size, is also the commonest of our species where it occurs, and it has a wide range over the Empire; but it is not found at higher elevations than 6,000 feet, nor in Ceylon or the extreme south of India. To the drier portions of the country it appears only to come in the monsoon. It has the characteristic habits of hemipodes to perfection, sitting particularly close in the low cover it affects, and when raised taking an even shorter flight than the other species, so that it can hardly be shot; while after this effort it sits so very tight that not only do dogs pick it up, but it has even been caught by hand.

In disposition it is about the tamest bird in existence; in a cage it will let one pick it up like a white mouse, and seems equally at home in close captivity, so that a pair of these tiny beings would make interesting pets for any one who likes birds, but can only find room for quite a small portable cage. In England they have even been known to lay in a cage, and at this time even threatened to charge the hand of their owner! What such tiny things could do against anything bigger than a mouse or a locust is a problem, but evidently they are not wanting in pluck. They have been found to feed on grass-seed and white ants, and are to be seen in gardens as well as in the open country. They are often found on land which has been flooded during the rains. The nest, sometimes domed and sometimes open, may be found even as late as October, in some places, though breeding begins as early as April; the eggs tend to be more numerous than in the larger species, for five and six may be found, though the usual hemipode clutch of four is more general. They are of the pointed peg-top shape, and show the typical dark peppering and spotting of the family; but are not so much smaller than those of the larger button-quails, as would be expected—another point of resemblance to the jack-snipe. The note is described as a "plaintive moan" or "a mixture of a purr and a coo," the bird when calling raising its feathers and turning about like a courting pigeon. This tiny bird is the smallest of our game birds, but, like the tallest, the sarus crane, is rather a bird for the aviculturist than the sportsman; if one wants to eat small birds, larks would be more worth shooting both for sport and for eating purposes. Besides the name *Dabki*, which means "squatter," this little bird rejoices in several others —*Turra*, *Libbia* and *Chimnaj*, in Hindustani; *Tella dabbagundlu* in Telugu, and *Darwi* at Ratnagiri. Yet we are told that natives, unless professional bird-catchers, generally consider it simply as a young quail of sorts, and certainly it has all the appearance of a young bird which ought to grow up into something quite different.

Nicobar Megapode.

Megapodius nicobariensis.

"Megapode" means big foot, and our single species, like Hercules, can be identified by its foot only, though, as it only inhabits the Nicobars, and the only other game-bird there is the local race of yellow-legged button-quail, which is neither big in body nor in foot, there is not much likelihood of anyone getting it often or mistaking it for anything else.

TURNIX ALBIVENTRIS

The bird itself is about as big as a jungle hen, and has the sides of the head red and bare like a fowl's, but its very short tail gives it rather the appearance of a guinea-fowl. Its plumage is unique among our game-birds by its very dulness, there being not a single streak or spot to relieve its monotony of snuffy-brown. The sexes are alike, and even the chicks hardly differ except by having downy heads. It is about all they do have downy, for they come out of the egg full-fledged, as is the usual custom of birds of the megapode family; their habits are well known in Australia, where not only a similar bird to this, locally called "jungle-fowl," but others of more distinct and handsome appearance, the "brush-turkey" (*Catheturus lathami*), and "mallee-bird" (*Leipoa ocellata*) are found. The type, indeed, is an Australian one, but the typical *Megapodius* group ranges east and west among the islands, ours being the farthest outlier to the westward.

The foot of the megapode has the hind-toe well developed, and furnished, like all the other toes, with a long, strong claw. It is thus better fitted for grasping than that of our other game-birds, and this power is employed by the bird in throwing up the great mounds in which its eggs are to be buried, for another queer habit of the family is to construct natural incubators for their eggs, which are of extraordinary size, in this species being as big as those of a goose, while the bird itself averages about a pound and a half in weight. Fresh eggs are ruddy pink, but they fade to buff as incubation advances, and also show white spots and streaks, caused by the colour, which is only a thin surface coating, getting chipped or scratched off.

The mounds are almost invariably situated just where the jungle abuts on the coral beach, not in the open, and very rarely back in the forest. Forest mounds are necessarily made of leaves and sticks mixed with earth; but evidently the proper compost, from the birds' point of view, is the coral sand of the beach, raked in a layer about a foot thick over a liberal foundation of leaves, cocoanut husks, and any sort of vegetable matter that these birds can lay their claws on. The same mounds are used again and again, the birds apparently scraping the top-dressing of sand off every now and then, putting on more vegetable refuse, and then raking the sand over again.

In this way an old mound may, although the Nicobarese say it is all the work of one pair, attain a height of eight feet and a circumference of sixty; but the mound of this size recorded by Davison as quoted by Hume was exceptional, and no doubt old, as it had a good-sized tree growing in it; about half the above dimensions represent the usual size.

In these mounds, at a depth of over a yard, the old bird buries her eggs, which hatch in the damp warmth generated by the decaying vegetation, aided no doubt by the lime in the coral and shell-sand. At the same time, the eggs will hatch when removed from the mound and left lying about anyhow; the young need no "mothering," but look after themselves from the first, and might easily be taken for some funny sort of quail. It is most likely that the old birds dig them out when due to hatch, for burrowing up through several feet of compost would be rather a heavy task even for a megapode chick, and the brush-turkey, which frequently breeds in zoological gardens, certainly digs the young out when due—in its case after six weeks' incubation.

The mound is thrown up at night—in fact, the bird is nocturnal altogether, and does not leave the shelter of the jungle in the day-time, while even at night it is the beach, and not the grass-land inland of the jungle-belt which it frequents. Although mostly a ground-bird, it often alights in a tree, and flies like a jungle-fowl. Its note is also like the cackling of a hen. These birds are a most valuable game-bird; they are abundant, being found often in flocks as well as pairs, give much the same sort of sport as jungle-fowl, and are, according to Hume, who was very critical about birds' table qualities, exceptionally good on the table, being both fat and succulent. Their food appears to be mostly animal, consisting of grubs and small snails; in captivity the young thrive on white ants.

Since as many as twenty eggs can be taken out of a mound and the young are easily reared, the Government should surely be approached with a view to disseminating this valuable bird all over our tropical islands where natural conditions are at all favourable.

F. Waller Chromo Lith. 38 Hatton Garden London

2/3

TURNIX DUSSUMIERI

INDEX.